The genus *Stachytarpheta* (*Verbenaceae*) in Brazil

Sandy Atkins[1]

Summary. The genus *Stachytarpheta* Vahl is reviewed as it occurs in Brazil. A total of 79 species is recognised, five of these with subspecies or varieties. Sixteen species and nine subspecies or varieties are described as new; five new combinations are made. The morphology is described, and characters used in a cladistic analysis are discussed. An infrageneric system dividing the genus into 12 informal groups is also proposed. Illustrations and keys are provided.

Resumo. O gênero *Stachytarpheta* Vahl foi revisado para o Brasil e está representado por 79 espécies, incluindo táxons infra-específicos (sub-espécies e variedades). Das 79 espécies reconhecidas para o gênero, 16 são inéditas, as quais são aqui descritas, e cinco novas combinações são propostas. O trabalho apresenta descrição morfologica e discussão dos caracters utilizados para a análise cladistica. Também está sendo proposto um sistema de classificação infra-genérico, dividindo o gênero em 12 grupos informais. Illustrações e chaves dicotômicas também são fornecidas.

Key words. *Verbenaceae*, *Stachytarpheta*, Brazil, taxonomy, morphology, new taxa.

Contents

Introduction

The aim of this work is primarily to describe all the species of *Stachytarpheta* occurring in Brazil. *Stachytarpheta* is a genus of about 133 species, (see below), almost exclusively from the New World (only *S. indica* (L.) Vahl, occurs in the Old World alone). The distribution in tropical and subtropical America is: Peru seven species; Ecuador five; Colombia 10; Venezuela seven; Guianas eight; Paraguay seven; Chile 0; Argentina five; Central America 24 (from herbarium data). In Brazil there are 79 species. It became obvious, while studying the Brazilian species, that it was necessary to look again at the infrageneric classification. Schauer (1847) was the first to propose such a classification, followed by Briquet in 1895. Since then many new species have been described, and the classifications have become inadequate both in terms of reflecting the variation shown by these new species, and in their relationships.

Accepted for publication February 2005.

[1] Herbarium, Royal Botanic Gardens, Kew, Richmond, Surrey TW9 3AB, U.K.

The species described are distinct from each other and easily recognised on gross morphological characters, as outlined in the morphological analysis. I have also used 2 different subspecific ranks. Where a group of individuals are linked together by obvious and important characters, such as inflorescence length and width, calyx morphology, position of anthers, fruit morphology (see results of analysis), but where there may be a large variation in less obvious or important characters, such as denseness of indumentum, leaf shape and size or bract shape and size, and where the distribution of these characters is geographically distinct, then I have assigned these groups to the rank of subspecies. Where these variations occur sympatrically, with overlap in the distribution of the different groups; then I have assigned the rank of variety.

Material and methods

The Kew herbarium has received thousands of specimens collected in Brazil, from joint expeditions with the Universidade de São Paulo (SPF), the Centro de Pesquisas do Cacau, Itabuna, Bahia (CEPEC), and more recently with the Universidade Federal de Bahia, Salvador (ALCB) and the Universidade Federal de Feira de Santana, Bahia (HUEFS). One of the main areas of joint research has been the investigation of the biodiversity of the Espinhaço range of mountains, particularly in the habitats known as *campo rupestre* (Harley & Simmons 1986; Giulietti *et al.* 1987; Stannard 1995; Zappi *et al.* 2003). Kew has also been the recipient of many duplicate specimens from Brazilian Institutions, particularly the Instituto de Botânica, São Paulo (SP) and the Museu Botânico Municipal, Curitiba, Paraná (MBM). I have been able to make three collecting trips to Brazil with the aid and collaboration of Brazilian colleagues at SPF, ALCB and HUEFS.

These specimens have revealed remarkable species diversity in *Stachytarpheta*, particularly in the *campo rupestre* areas of Bahia and Minas Gerais. As well as the very rich collections at Kew, the author has studied material sent on loan from the following institutions: BM, BR, E, G, HUEFS, NY, RSA, S, TEX, and SPF. Only a selection of the examined collections is cited after each taxon. A complete list of collections is given in Appendix C. All the cited types, including synonyms, have been seen by the author unless the citation is followed by 'n.v'.

J. C. Schauer (1847, 1851) wrote the accounts for *Verbenaceae* in de Candolle's *Prodromus* and Martius' *Flora Brasiliensis*. It appears that he wrote them simultaneously. From correspondence in the archives at the Geneva herbarium, he started in 1842 while he was director of the Botanical Garden in Breslau, and finished after he had moved to be Professor of Botany at the University of Greifswald in 1844. From the correspondence it can be seen that he worked on specimens in these places, and it is to be presumed that the specimens were sent to him from Geneva and Munich. Unfortunately he does not cite specific types, but refers to a few specimens he has seen for each species, often not giving collectors' numbers. The de Candolle herbarium is still intact in Geneva, and many specimens carry Schauer's own determinations, as do some of the very good set of *Flora Brasiliensis* duplicates in BR. I have assumed that where Schauer's own handwriting is on the sheet, that the specimen is a probable syntype, if it carries no other possible means of certainty, e.g. collector's number or exact locality. I have chosen not to lectotypify specimens here unless there is a confusion of concepts.

A morphological cladistic analysis was carried out to provide a classification that is more consistent with morphological variation and hypothetical relationships.

The species descriptions and a morphological matrix for cladistic analysis have been largely compiled from dried material. The methods of the cladistic analysis are presented in the morphological analysis section.

History

Stachytarpheta was first described by Vahl (1804) with 12 species. Vahl erected the genus to accommodate species which had been described as verbenas, but which possessed only two fertile stamens and two staminodes, and had fruit that divided into two, not four mericarps. Link (1821) misspelt the name as "*Stachytarpha*", this spelling was taken up by Schauer (1847). Pohl (1827), who travelled to Brazil with Spix and Martius between 1811 and 1817, erected a genus *Melasanthus* in which he described six species, collected by him, all with black or dark purple corollas. He put the new genus in *Verbenaceae*, but thought the nearest relative was *Schwenkia* in the *Scrophulariaceae*. His drawings and specimens leave no doubt that they are species of *Stachytarpheta*. By 1847, Schauer had had access to very many collections from Brazil, including those of Sellow, Martius, and Pohl. His account for de Candolle's *Prodromus* contained descriptions of 43 species, of which 34 were described as occurring in Brazil. He was the first to subdivide the genus. He distinguished two sections, *Abena* (sect. *Stachytarpheta*) and *Tarphostachys*, based mainly on the thickness of the spike, depressions in the rachis, shape of corolla, length of calyx and corolla tube, and exsertion of the style above the corolla tube. He further subdivided section *Abena* into two subsections, *Lepturae* and *Pachyurae*, and section *Tarphostachys* into 4 subsections, *Longispicatae*, *Brevispicatae*, *Subspicatae* and *Capitatae*.

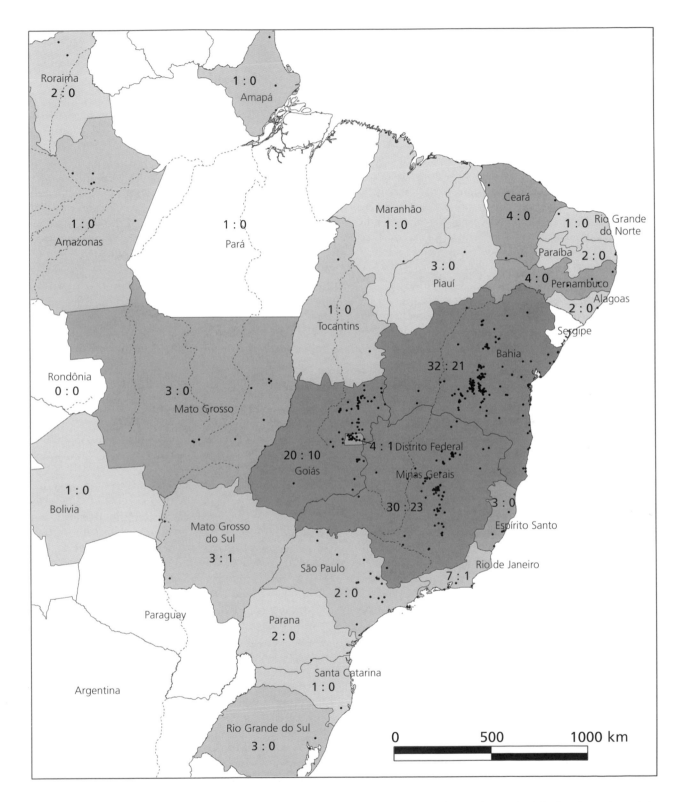

Map 1. Map showing levels of endemism of *Stachytarpheta* within each state in Brazil. The number on the left of each number pair is the number of species of *Stachytarpheta* occurring in that state, the number on the right of each number pair is the number of endemic species to that state.

Briquet (1895) divided the genus into two sections, *Abena* and *Melasanthus* (section *Tarphostachys* Schauer). It is not clear why Briquet renamed *Tarphostachys*. Pohl used *Melasanthus* to name a new genus based on six species from Goiás, while Briquet used it to re-name Schauer's section *Tarphostachys* which contained 26 species, including Pohl's six species. Briquet may have considered Pohl's *Melasanthus* (1828) to have priority over Schauer's of 1847 although a name only has priority at its own rank. This supposition is supported by his taking up of Schauer's subsections. Briquet's classification was based on the arrangement of flowers and bracts in a spike, angular or terete rachis, size of depressions in the rachis and presence or absence of scale-like bracts. Moldenke (1959a, 1971) adopted the same sections and subsections even though he described many new species.

Distribution and biogeography

My estimate for the total number of species worldwide is 133. This excludes the hybrids produced by and given names by Danser (1929). Five species purport to have their origin outside the New World: *Stachytarpheta angolensis* Moldenke, *S. bogoriensis* Zoll. & Moritzi, *S. fallax* A. E. Gonç., *S. jabassensis* H. Winkl., and *S. trimenii* Rech. The type specimen of *S. angolensis* is conspecific with previously unidentified specimens in the Kew herbarium from Central America. *S. bogoriensis* has been put into the synonymy of *S. jamaicensis* (Munir 1992). *S. fallax* (described from The Cape Verde Islands), appears to be *S. cayennensis*. *S. jabassensis* from west tropical Africa was collected on waste ground close to a factory, and although I have not seen the type, I suspect from the description, and the locality, that it also is *S. cayennensis*. *S. trimenii*, reported from Kandy in Sri Lanka, is a natural hybrid between *S. indica* L. (see below) and *S. mutabilis* (Jacq.) Vahl (from Tropical America, widely naturalised in all tropical areas).

Stachytarpheta is therefore almost a wholly New World genus, some species of which have spread and become naturalised. The only exception is *S. indica*, which seems to have no New World distribution. According to Verdcourt (1992), this is a widespread African species. It seems to be closely related to *S. angustifolia* from the New World. It could be that a very early introduction from the New World to Africa of *S. angustifolia* has had time to evolve a different morphology, possibly as a result of the "Founder Effect" where populations started from a small number of individuals, isolated from the parental population, result in a reduction of the gene pool and alteration of gene frequencies (Ridley 1996). (See notes under *S. angustifolia*).

Stachytarpheta jamaicensis, *S. cayennensis* and *S. mutabilis* have become widespread in tropical and subtropical regions. Within the tropics and subtropics of America, the areas of greatest species diversity are Mexico and Brazil. Mexico has 22 species, of which 14 are endemic (Moldenke 1971). Brazil has 79 species, of which 73 are endemic (Table 1). The distribution shows no overlap of taxa between the mountains of Brazil and Venezuela except for *S. sprucei*, which is found in the border area between Venezuela, Brazil and Guyana.

Table 1. Showing numbers of species and numbers of endemics of *Stachytarpheta* occurring in Latin American Countries

Country	Number of species	Number of endemics
Mexico	22	14
Central America	24	1
Colombia	10	0
Venezuela	7	1
The Guianas	8	1
Ecuador	5	2
Peru	7	4
Bolivia	4	0
Chile	0	0
Paraguay	7	2
Argentina	5	2
Brazil	79	73

Within Brazil, most of the species are narrow endemics concentrated within the states of Goiás, Minas Gerais and Bahia. These three states have been the subject of many botanical expeditions because of their high levels of endemism in a number of groups (Giulietti *et al.* 1987; Giulietti & Pirani 1988, 1997; Harley 1988, 1995; Harley & Simmons 1986; Stannard 1995; Zappi *et al.* 2003). The levels of endemism for *Stachytarpheta* are shown in Map 1. The figures revealed on the map show a high level of endemism. There is virtually no overlap of species between the states of Bahia, Goiás and Minas Gerais (Table 2), and *Stachytarpheta* occurs in most of the distinct vegetation types in these areas. The species which can be found in other countries in South America (Table 1) are: *S. angustifolia*, *S. cayennensis*, *S. jamaicensis*, *S. paraguariensis*, *S. polyura*, and *S. sprucei*.

The Serra do Espinhaço, a centre of diversity of *Stachytarpheta* with well over half the species found here, forms an elongated mountain range over 1000 km long from north to south (Harley 1995), between 10°00'S, near the Serra da Jacobina in Bahia, to 21°25'S in Minas Gerais, near Ouro Preto. Much has already been written about the geology (Harley 1988; Giulietti & Pirani 1988), the climate (Galvão & Nimer

Table 2. Species as distributed by state as cited in this account
* denotes endemic to the state

ACRE	S. cayennensis
ALAGOAS	S. angustifolia, S. hirsutissima
AMAPÁ	S. angustifolia
AMAZONAS	S. cayennensis
BAHIA	S. almasensis*, S. angustifolia, S. arenaria*, S. bicolor*, S. bromleyana*, S. caatingensis*, S. cayennensis, S. coccinea, S. crassifolia subsp. crassifolia*, S. crassifolia subsp. abairensis*, S. crassifolia subsp. rotundifolia*, S. froesii*, S. galactea*, S. ganevii*, S. glandulosa*, S. guedesii*, S. hatschbachii*, S. hirsutissima, S. hispida*, S. jamaicensis, S. lactea, S. lacunosa*, S. lychnitis*, S. lythrophylla, S. macedoi, S. maximiliani, S. microphylla, S. piranii*, S. quadrangula*, S. radlkoferiana*, S. radlkoferiana var. lanata*, S. scaberrima, S. stannardii*, S. trispicata*, S. tuberculata*
CEARÁ	S. angustifolia, S. cearensis, S. coccinea, S. sessilis
DISTRITO FEDERAL	S. gesnerioides, S. lactea, S. longispicata subsp. ratteri*, S. polyura, S. villosa
ESPÍRITO SANTO	S. hirsutissima, S. restingensis, S. scaberrima
GOIÁS	S. angustifolia, S. atriflora*, S. candida*, S. cayennensis, S. confertifolia, S. dawsonii*, S. gesnerioides, S. gesnerioides var. glabra, S. glauca, S. glazioviana*, S. integrifolia*, S. lactea, S. longispicata subsp. longispicata*, S. longispicata var. andersonii*, S. longispicata var. longipedicellata*, S. longispicata var. parvifolia*, S. macedoi, S. martiana, S. pachystachya, S. polyura, S. puberula*, S. rhomboidalis*, S. sericea*, S. sericea × longispicata*, S. villosa
MARANHÃO	S. sessilis
MATO GROSSO	S. gesnerioides, S. gesnerioides var. glabra, S. gesnerioides var. hirsuta, S. paraguariensis
MATO GROSSO DO SUL	S. matogrossensis*, S. paraguariensis, S. polyura
MINAS GERAIS	S. ajugifolia*, S. alata*, S. amplexicaulis*, S. brasiliensis*, S. cassiae*, S. coccinea, S. commutata*, S. confertifolia, S. crassifolia subsp. minasensis*, S. diamantinensis*, S. discolor*, S. glabra*, S. harleyi*, S. itambensis*, S. linearis*, S. longispicata subsp. brevibracteata*, S. longispicata subsp. minasensis*, S. macedoi, S. martiana, S. mexiae*, S. monachinoi*, S. pachystachya, S. pohliana*, S. procumbens*, S. reticulata*, S. rupestris*, S. scaberrima, S. sellowiana*, S. spathulata subsp. mogolensis*, S. spathulata subsp. spathulata*, S. speciosa, S. viscidula*
PARÁ	S. angustifolia
PARAÍBA	S. angustifolia, S. maximiliani
PARANÁ	S. cayennensis, S. polyura
PERNAMBUCO	S. angustifolia, S. cayennensis, S. cearensis, S. jamaicensis
PIAUÍ	S. lythrophylla, S. microphylla, S. pachystachya
RIO DE JANEIRO	S. angustifolia, S. cayennensis, S. hirsutissima, S. laevis, S. restingensis, S. schottiana*, S. speciosa
RIO GRANDE DO NORTE	S. sessilis
RIO GRANDE DO SUL	S. cayennensis, S. laevis, S. matogrossensis, S. sessilis
RONDÔNIA	
RORAIMA	S. jamaicensis, S. sprucei
SANTA CATARINA	S. paraguariensis
SÃO PAULO	S. cayennensis, S. gesnerioides var. gesnerioides
SERGIPE	
TOCANTINS	S. macedoi

1965) and the biodiversity (Giulietti & Pirani 1988; Harley 1988, 1995; Zappi *et al.* 2003) of this area.

By and large, the species of *Stachytarpheta* are narrow endemics. There are some links between Goiás and Minas Gerais, e.g. *S. martiana*, to be found in localities some 700 km apart and separated by the great valley created by the São Francisco river; *S.*

confertifolia, with a locality just south of Belo Horizonte and another in the Chapada dos Veadeiros. This latter distribution could relate to the ancient occupation (dating to the Cretaceous) of the *cerrado* vegetation of nearly 2 million km² of the land surface of Brazil (Ratter & Dargie 1992), which has since become fragmented. The fairly common and widespread species of the *cerrados* of Goiás and Distrito Federal, *S. longispicata*, has outlying subspecies in Minas Gerais.

The links between Minas Gerais and Bahia are confined to two lowland species, *Stachytarpheta coccinea* and *S. scaberrima*, with localities associated with *caatinga* or disturbed *caatinga*. All other connections between these three states (Bahia, Goiás and Minas Gerais) involve common and widespread species.

The five species associated with coastal *restinga* appear to have no other links; their localities are confined to the coastal areas. Some of these, like *Stachytarpheta schottiana* are found in a small area around Rio de Janeiro only, but *S. maximiliani* is found from Paraíba down to the south of Bahia, and *S. hirsutissima* is found in Alagoas, Bahia, Espírito Santo and Rio de Janeiro.

As stated, the majority of the species are very narrowly distributed. The Chapada Diamantina National Park contains six of the Bahian endemics. There are other very important areas within Bahia containing endemic species, especially areas around Rio de Contas, Catolés, Barra da Estiva and Morro do Chapéu. Examples include *Stachytarpheta bromleyana*, collected only once, from the Serra do Ouro near Barra da Estiva; *S. stannardii* from the area around Tijuquinho, Mun. Abaíra; *S. arenaria* from Serra dos Brejões, Serra dos Brenhos and the Serra dos Cristais, Mun. Abaíra and Rio de Contas; *S. ganevii* from Riacho das Taquaras and Porco Gordo, Mun. Abaíra and Rio de Contas.

In Minas Gerais, the examples of localised distribution are also numerous: *Stachytarpheta ajugifolia* and *S. sellowiana* from the Serra de São José outside Tiradentes; *S. commutata* from Mount Itacolomi near Conceição do Mato Dentro; *S. diamantinensis* collected only once from near Diamantina; *S. discolor* from around Serro; *S. itambensis* from the Pico do Itambé; *S. linearis* from around Conselheiro Mata; *S. monachinoi* from the Serra do Cabral; *S. procumbens* from the Serra do Cipo; from further north in the state, *S. cassiae* from the Serra do Pau Dárco, near Espinosa, and others with only slightly wider distributions.

This distribution pattern has already been highlighted in other quite unrelated plant groups (Giulietti & Pirani 1988; Harley 1988; Zappi *et al.* 2003), and suggests explosive speciation and adaptive radiation, some of it sympatric.

Conservation

Conservation assessments using accepted IUCN categories (IUCN 2001) have been given after each species description. Many of the species of *Stachytarpheta* are restricted to particular habitats, and as already stated, the majority of species are found in the *campo rupestre* areas. Many of these areas are under threat from encroaching agriculture. At higher altitudes, where the soil is shallow, only subsistence farming occurs, but in a few places, on more extensive areas of level ground and where water is available, cash crops such as coffee are planted. Cattle farming is also an important activity, and destruction of these micro-habitats is caused by trampling, eutrophication of soils (depositing of faeces rich in nitrogen and phosphate), and the annual burning carried out by farmers to promote regrowth of herbaceous, particularly grassland vegetation.

The widespread liming of *cerrado* soils, which are then used for soya bean cultivation has already accounted for the destruction of huge areas of *cerrado* (Harley *pers. comm.*), and forest and *cerrado* areas are used for fuelwood for domestic use, and in the production of charcoal (Giulietti & Pirani 1997).

In large areas of Minas Gerais, local vegetation is being cleared for *Eucalyptus* plantations, and in Bahia many areas of forest have been cut down to provide sites for coffee plantations (Harley *pers. comm.*).

The *restinga* species are especially at risk from large-scale building along the dune areas behind the beaches and consequent urbanisation. This urbanisation caused by population increase is also a threat in Distrito Federal. Brasilia was projected to accommodate 500,000 people by the year 2000, but the population is now estimated at 2,000,000. This causes problems of water shortage, and encroachment by building caused by chronic housing shortage (Filgueiras 1997).

However, the creation of the National Parks of Chapada Diamantina, Serra do Cipó, Serra da Canastra, Monte Pascoal, Brasília and Chapada dos Veadeiros as well as other smaller protected areas, have all helped to conserve many localities for *Stachytarpheta*.

Affinities

Schauer (1847) placed *Stachytarpheta* in tribe *Verbeneae*, subtribe *Verbeneae*, along with *Verbena*, *Bouchea* and *Lippia*, but not *Lantana*. He based this mainly on the fruits being schizocarpic rather than drupaceous. Briquet (1895) placed *Stachytarpheta* in tribe *Lantaneae*. The characteristics of this tribe are: fruit fleshy with 2-locular, 2-seeded pyrene, or a dry or sub-fleshy schizocarp separating at maturity into two 1-seeded mericarps. Junell (1934), however, showed that convergent abortive reduction of the abaxial carpel has occurred several times within the family. He implied that *Stachytarpheta*, *Bouchea* and related genera with only one functional carpel would be better placed in the *Verbeneae*. Sanders (2001) also suggests that these genera should be included within *Verbeneae* because of such putative apomorphies as inflorescence structure, calyx and corolla shape, structure of the anthers and connectives, staminode presence, style-branch orientation and fruit structure which indicate that they have more in common with *Verbena,* with its dry schizocarp, separating at maturity into four 1-seeded mericarps.

Stachytarpheta is most closely allied to *Chascanum* and *Bouchea* (see cladistic analysis below). These three genera all have an inflorescence which is a terminal spike, all have calyces which are narrow and 5-toothed (basically 5-toothed, but with variations), filamentous styles and schizocarpic, narrowly cylindric fruits with two mericarps. *Stachytarpheta* is separated from *Bouchea* and *Chascanum* by having only two fertile stamens as opposed to four, the posterior pair being replaced by staminodes; by its style branches fused into a discoid head, completely stigmatic as opposed to the style lobes forming a hook, with the stigmatic part oblique; and with anther thecae divergent, as opposed to parallel. *Chascanum* is a genus of about 27 species and occurs in Africa, Madagascar, Arabia and India. *Bouchea* is a genus of about nine species and occurs in tropical and subtropical America.

Morphology of Brazilian species

Life form and habit

The most common form in the genus is a branched shrub or sub-shrub, normally 1 – 1.5 m high, although *Stachytarpheta coccinea, S. quadrangula* and *S. trispicata* attain 2 – 3 m in height, and are more like treelets. *S. piranii* has a reported height of 4 m. Only *S. angustifolia, S. macedoi* and *S. lythrophylla* seem to be annuals and occur in marshes, and in seasonally flooded areas around rivers. *Stachytarpheta chapadensis,* and *S. sessilis,* are delicate herbs, reaching only 20 – 25 cm, but are believed to be perennial from a woody rootstock. Similarly, species which are known to grow in dry *cerrado* or open *campo geral* such as *S. longispicata, S. cabralensis, S. gesnerioides, S. linearis* and *S. sericea* produce several simple stems from the same kind of woody xylopodium. *S. stannardii* and *S. arenaria* are large decumbent shrubs, while *S. candida* and *S. procumbens* are perennial herbs which grow with horizontal stems hugging the ground, with only the inflorescence erect.

Stems and indumentum

Stems of *Stachytarpheta* can be rounded, round-quadrangular, or distinctly quadrangular. Indumentum is varied. Where a plant is hairy, it is generally more hairy towards the apex of the plant as the indumentum seems to be lost as the stem becomes woody with age. Sometimes the stem is hairy on opposite faces, while the other two faces are glabrous, or more hairy at the leaf nodes. Hairs can

be simple (i.e. single-celled), or uniseriate (i.e. formed from two or more cells in a line which is unusual in the family), and occasionally gland-tipped (Plate 9C, D), and can be erect, patent or pointing in all directions. Stems are rarely glandular.

Leaves

Leaves are sessile or petiolate; linear, elliptic, ovate, rhomboid, or spathulate, usually with crenate or serrate margins, but entire in some species. A common feature of the leaf base is that it is decurrent into the petiole. All leaf measurements are averages taken over several specimens, and include the petiole (i.e., the measurement is taken from the joint with the stem to the leaf apex because the petiole is often not distinct). The leaf arrangement is usually decussate, with leaves patent to the stem, but there are several species where the leaves are imbricate and erect.

Leaf surfaces are glabrous or variously hairy. The indumentum can be distributed just on the lower leaf surface, or on both surfaces, sometimes just along nerves. Hairs can be simple or uniseriate (Plate 9C, D), all pointing in same direction, or in all directions.

Some species have very conspicuous nectariferous glands on the leaves, bracts and calyces (Plate 9G, H). These may attract insects, particularly ants, as a protection against herbivores, particularly in the *cerrado* vegetation (Oliveira & Leitão-Filho 1987). However, there is evidence in *Stachytarpheta* that much of the nectar from the base of the ovary is robbed, without pollination taking place. In dried herbarium specimens, there are often minute holes at the base of the corolla, usually caused by short-tongued bees that cannot otherwise reach the floral nectar, and in some species, where this is prevalent, very little fertilisation has taken place. It is possible, therefore, that the glands are acting as a food source decoy, making nectar readily available to prevent the robbing from the reproductive zone.

The glands may be of different kinds (Plate 9G, H), and more work needs to be done on their anatomy.

Inflorescence

Inflorescence size (length and width) has been used as the basis for infrageneric classification by both Schauer and Briquet. The flowers, each subtended by a bract, are arranged spirally along the inflorescence axis, but are packed so tightly together that the arrangement and definition of a floral axis is difficult to determine. Flowers mature and open only 2 – 3 at a time, from the bottom of the inflorescence towards the top, with both fertilised and immature flowers together on the same spike. There are basically four different kinds of inflorescence:

(i) Long, narrow, whip-like inflorescences. This inflorescence type can be up to 54 cm long, with more than 35 flowering nodes.

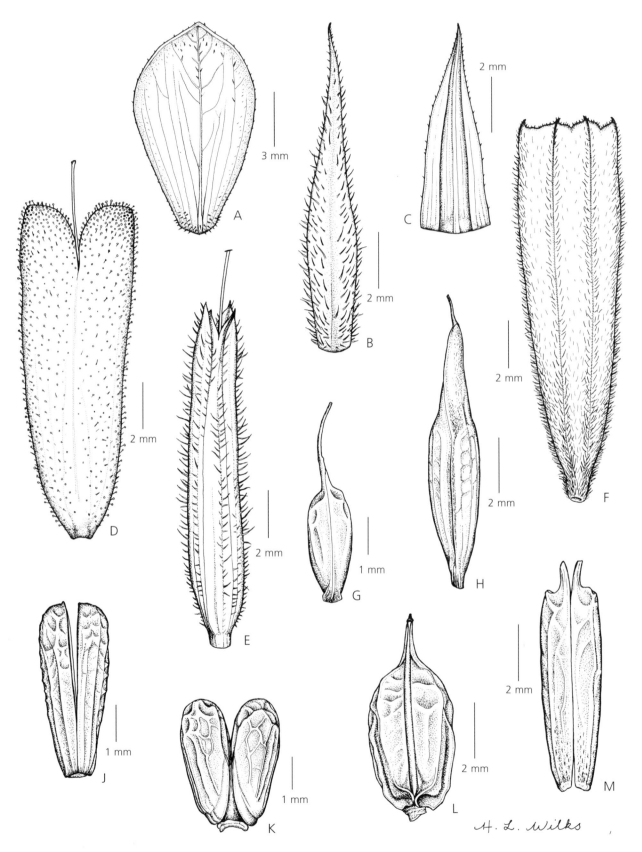

Fig. 1. Range of bracts, calyces and fruits in *Stachytarpheta*. **A – C** bracts. **A** *S. ganevii*; **B** *S. speciosa*; **C** *S. integrifolia*. **D – F** calyces. **D** *S. almasensis*; **E** *S. coccinea*; **F** *S. longispicata* subsp. *longispicata*. **G – M** fruits. **G** *S. cayennensis*; **H** *S. coccinea*; **J** *S. lacunosa*; **K** *S. ganevii*; **L** *S. martiana*; **M** *S. microphylla*. DRAWN BY HAZEL WILKS.

(ii) A slightly shorter, thicker inflorescence. This inflorescence type can be up to 17 cm long, with 20 – 26 flowering nodes.

(iii) Inflorescences with 6 – 12 flowering nodes, up to 10 cm long, but normally in the range of 4 – 7 cm, quite dense.

(iv) Inflorescence up to 3 cm long, with up to 4 flowering nodes. A compact crowded head, sometimes with the surrounding leaves almost overtopping the inflorescence.

Bracts

The floral bracts are herbaceous or woody, green, sometimes tinged purple, or red, and closely appressed against the calyx. They vary greatly in size and shape, and are sometimes keeled, but seldom exceed the calyx in length. They can be very small and inconspicuous, or large and brightly-coloured. They are often persistent, even after the fruit and calyx have dropped from the inflorescence. They can be glabrous, hairy, or ciliate along the margin (Fig. 1A – C).

Calyx

The calyx is tubular and narrow, usually more than half the length of the corolla tube; herbaceous, green, or tinged purple or red; sometimes ribbed, sometimes smooth, essentially 5-toothed, sometimes ± regular, but sometimes with fused teeth, and often with one or two sinuses in place of teeth (Plate 9E, F). The teeth are triangular or rounded. The calyx does not enlarge in fruit, but becomes brittle and papery and completely encloses the fruit. The calyx teeth often twist over the top of the fruit to form a seal, and then the calyx splits longitudinally to release the fruit. Occasionally, the top of the calyx remains open, but the calyx splits in the same way. Although Schauer (1847, 1851) in his descriptions, observed great differences in the arrangement of the calyx teeth, he did not use the character to group the species. There are three main arrangements of teeth. (Note: the term adaxial, when used to describe the calyx means the side held nearest to the inflorescence rachis, i.e. the inside face, and the term abaxial means the side furthest away from the inflorescence rachis, i.e. the outside face).

(i) There is no sinus. The five calyx teeth are ± regular; occasionally the adaxial tooth is slightly larger or smaller than the other four (Fig. 1F, Plate 9F).

(ii) There is a sinus on the adaxial side. This appears to take the place of the adaxial tooth. The remaining four teeth are on the abaxial side, and can be regular, or the outer two larger than the inner two (Fig. 1E).

(iii) There is a sinus on both the adaxial side and the abaxial side. In this case, the calyx seems bifid, with no teeth expressed, but occasionally minute teeth can be detected on each side (Fig. 1D, Plate 9E).

Corolla

Corollas are either infundibular or hypocrateriform, with a narrow tube and spreading lobes. The tube is sometimes kinked, and may exceed the calyx or may be ± equal to it. There are five lobes, either ± equal or with two adaxial lobes slightly smaller than the abaxial, rounded. Where a lobe measurement is given in the description, this refers to a mean of measurements taken across the broadest part of a single lobe, i.e. an upper lobe; where a limb measurement is given, this is the mean of measurements taken across the limb of the corolla at its widest point, i.e. across the horizontal median line. The corolla can be red, blue, mauve, white or black. The colours are intense, and can be seen from a great distance. There are some striking coincidences of flower colour, as in a group of five species, all growing in a small area in Goiás, all with black corollas, and another group in Bahia, with bright red corollas. These flower colours may have developed to attract localised pollinators.

The stamens in *Stachytarpheta* are included within the corolla tube. The stigma is sometimes held well down inside the tube, but sometimes is above or at the same level as the stamens. Nectar is produced in small quantities from a disk just below the ovary at the base of the corolla tube. The opening at the mouth of the corolla is often very small, sometimes less than 1 mm at its widest point. This suggests that the main pollinators are insects with long tongues, probing for the nectar at the bottom of the tube. The individual flowers have only a relatively small 'landing area' at the point where the lobes spread at the apex of the tube. Small *Hymenoptera*, *Diptera* and humming birds are frequent visitors (pers. obs.). Most pollination appears to be effected by small nectar-seeking bees (Atkins *et al.* 1996).

There are often glandular hairs on the upper outside of the tube, and hairs and/or sessile glands at the throat, and often a ring of hairs on the inside of the tube just above the ovary. After fertilisation, the flower quickly fades and falls.

Androecium

The androecium consists of a single pair of fertile stamens, plus two (abaxial) staminodes, arranged didynamously. The filaments are rounded or flattened, glabrous or pubescent, normally very short and epipetalous, with anthers almost sessile. Depending on the point of attachment of the filaments, the anthers are either held well down inside the tube, or just at the throat. The anthers are dorsifixed, with the thecae strongly divaricate, which is unique in the family. There is sometimes a thickening of the connective.

Pollen

The pollen of *Stachytarpheta* is 3-colpate, with an exine sculpture made up of verrucae, which distinguishes it from all other genera in the family (Atkins 2004). According to Raj (1983) the pollen grains are peroblate to oblate with an equatorial diameter of 96 – 155 µm (pollen is normally visible to the naked eye). Atkins (1991) found grains of *S. sericea* with an equatorial diameter up to 175 µm. The pollen wall is between 5 – 10 µm thick (Fig. 2).

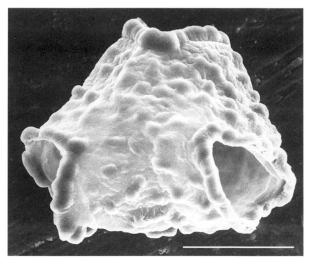

Fig. 2. Pollen. *Stachytarpheta integrifolia.* (SEM). Scale bar = 20 µm.

Gynoecium

The ovary is superior, syncarpous, ovoid, pear-shaped or flask-shaped, glabrous, normally about 2 mm long; the ovules are erect, fixed at the base of the locule. It sits on a disk, which is sometimes thin, sometimes thicker and slightly fleshy. The apex of the ovary is sometimes narrowed. The style is filamentous, with the style branches fused into a stigmatic, capitate head.

The fruit is a schizocarp, light or dark brown or black, ovoid, oblong, or narrowly cylindric. The outer surface is normally reticulate, but in some species is smooth, and in some of these species, the smooth fruit is covered with sessile glands. The commissure is normally hardly discernible as a fine line between the locules, but in some species it is a raised and prominent ridge. The base of the fruit is sometimes ragged where it has torn away from the disk at the base of the ovary, leaving a prominent scar, but sometimes comes away cleanly, leaving no scar (Plate 9A, B).

The abscission point at the base of the style sometimes occurs above the apex of the ovary, leaving behind a short stump or stylopodium, and sometimes the abscission point is below the level of the apex of the ovary, and comes away at fruit maturity. In some species the apex of the fruit is elongated into a short beak (Fig. 1G – M).

Morphological Analysis

Constructing a data matrix for a morphological cladistic analysis at species level has proved difficult. The synapomorphic characters used to define the genus, (only two fertile stamens, anther thecae divergent, style capitate) are of no value in defining relationships between species. The identifying characters which are useful in recognising species, e.g. inflorescence length and width, leaf indumentum, flower shape and size, are mostly quantitative. As far as possible, discreet qualitative characters were used, but, some quantitative characters were included, and the coding has been as explicit as the character allows.

In all, 34 characters were scored, (Appendix A) and 56 taxa were sampled (Appendix B). The taxa were chosen from across the morphological range of the genus, and included at least one species from each of Schauer's original groups. It is inescapable that nearly all of these species are Brazilian endemics. However, *Stachytarpheta jamaicensis*, a pantropical species, and two non-Brazilian species, *S. frantzii* Polak and *S. miniacea* Moldenke from Central America were included, as were *Chascanum laeta* and *Bouchea agrestis*.

Heuristic analyses were carried out using PAUP* vers. 4.0 software (Swofford 2002) on a Macintosh G4. Searches were conducted using Fitch (1971) parsimony with equally weighted characters, TBR (tree-bisection-reconnection) branch-swapping, and random taxon additions (1000 replicates) with the MulTrees option in effect. Only ten trees were saved per replicate to avoid extreme swapping on suboptimal islands. Internal support for clades was estimated using 100 bootstrap replicates (Felsenstein 1985), with simple taxon addition, TBR branch-swapping, and the MulTrees option in effect, holding ten trees per step.

Results

In this analysis 456 equally parsimonious trees were recovered, each of length 214 steps, with CI = 0.19, HI = 0.8 and RI = 0.6.

The strict consensus tree (Fig. 3) is shown with bootstrap values added. As previously stated, *Bouchea* and *Chascanum* are thought to be closely related to *Stachytarpheta*, and characters 29, 31 and 33 (see Appendix A) serve as generic synapomorphies supporting the monophyly of *Stachytarpheta*.

The ingroup shows little resolution. There are eight clades, four of which have bootstrap values > 50%.

Clade I is a group of species from Goiás, excluding *Stachytarpheta lacunosa*, which share black corolla, a 5-toothed regular calyx and an open fruiting calyx. The inclusion of *S. lacunosa* is supported by a woody rootstock and other characters which are homoplaseous in the clade.

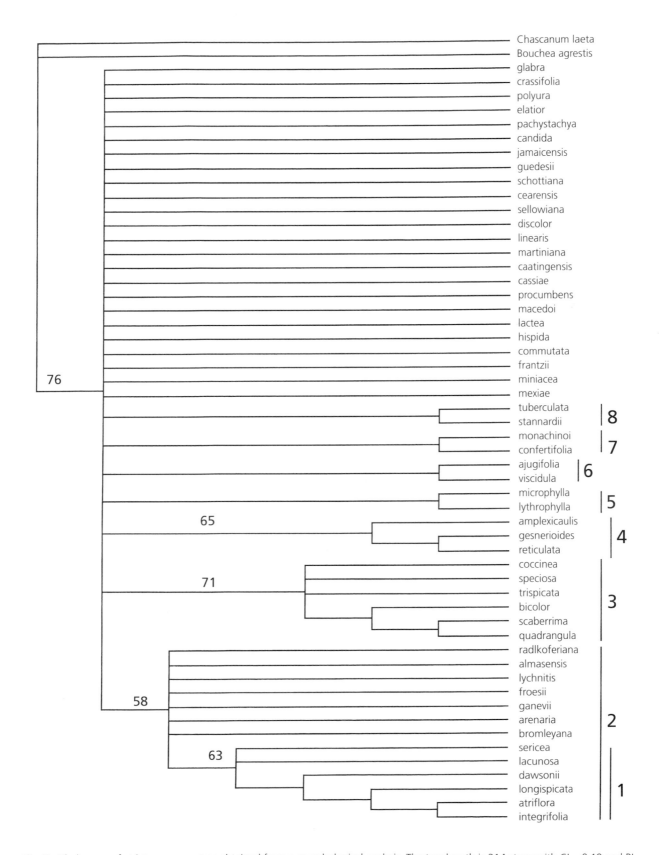

Fig. 3. Cladogram of strict consensus tree obtained from a morphological analysis. The tree length is 214 steps with CI = 0.19 and RI = 0.6. Bootstrap percentages are shown above the branches.

Clade II, includes Clade I plus *Stachytarpheta radlkoferiana, S. almasensis, S. lychnitis, S. froesii, S. ganevii, S. arenaria* and *S. bromleyana.* All these species are Bahian endemics. They share a bifid calyx, red corolla and a fruit without a stylopodium. However, these characters are not shared with the species of Clade I.

Clade III has a bootstrap value of 71%. These species share a growth habit of tall shrub or treelet, a bifid calyx, a long narrow corolla with small lobes and a smooth, beaked fruit.

Clade IV has a bootstrap value of 65% and the species share a woody rootstock, long, narrow inflorescence, pale reticulate fruit with apex flattened with short stylopodium.

Clades V – VIII are on the strict consensus tree but have bootstrap values less than 50%. Clade V, *Stachytarpheta microphylla* and *S. lythrophylla* share almost all the characters except habit, corolla colour, corolla tube length and lobe width.

Again, Clade VI, *Stachytarpheta viscidula* and *S. ajugifolia* are similar in all characters.

In Clade VII, *Stachytarpheta monachinoi* and *S. confertifolia* are similar in all characters except corolla tube length and position of anthers.

In Clade VIII, *Stachytarpheta tuberculata* and *S. stannardii* are similar in all characters except type of hair and inflorescence length.

Discussion

It is not possible to make detailed evolutionary statements on the basis of this morphological analysis. Lack of resolution in the strict consensus is probably a result of homoplasy in a number of characters. The presence of a woody rootstock, the types of hairs, the presence or lack of extra floral nectaries seem to show reversal or parallelism in different parts of the tree. Non-homoplasious characters are fruit shape and calyx shape.

It appears that microgeographic speciation has occurred, possibly by a process of recolonisation of a changing habitat caused by partitioned mountains, local rainshadows or other climatic effects causing shifting vegetational zones etc. There is evidence that this kind of speciation frequently involves polyploidy, hybridisation or cleistogamy (Gentry 1982). *Stachytarpheta* can hybridise in the wild (Atkins 1991), and from personal observations, it is possible that pollination can take place without an external agent. Unfortunately I have found no direct information on work on chromosome numbers. Munir (1992) cites Fedorov (1974) as reporting $2n = 48$ in *Stachytarpheta cayennensis* and $2n = 160$ in *S. indica*; and Sanders (2001) quotes $2n = 18, 48, 56, 112$ and 160 within the *S. cayennensis* complex.

Shifts in specific pollinators can be a common mode of speciation, and co-evolution can give rise to precise plant-pollinator systems; there are five species, all confined to the Chapada dos Veadeiros in Goiás, with black flowers. Harley (1986) published an account of a hybrid population of *Hyptis cruciformis* and *H. pachyphylla* from the same area, and made observations on local pollinators. In such groups speciation appears an open-ended phenomonon with no limits on species diversity.

I hope that molecular work will be undertaken at some point in the future; it may provide a more robust phylogeny. However, I want to provide a classification that groups species together using characters which are less variable in the genus and can be used to facilitate identification. The following taxonomy is based partly on the cladistic analysis as outlined, using, as far as possible, the non-homoplasious characters revealed by the analysis, but where the phylogeny is unresolved, it is not wholly compatible. The main aim of this work is to provide a workable and usable classification with groups which can be diagnosed by morphological characters.

The following 12 informal groups are presented. A key is provided below.

Group 1 Cayennensis. Subshrubs and herbs without a woody rootstock; inflorescence with up to 35 flowering nodes, tightly packed, narrow (up to 6 mm); bracts mostly woody; calyx erect, embedded in excavations in the inflorescence axis with 4 short teeth and a 5th tooth or sinus, or calyx appearing 2-toothed, because inner and outer tooth have merged or twisted together; corolla tube 5 – 20 mm, lobes 1 – 4 mm, hypocrateriform, mauve, blue or white; anthers at throat, or just below; fruit surface reticulate; apex with short stylopodium.

Nineteen species in Brazil.

Group 2 Gesnerioides. Subshrubs with a woody rootstock; inflorescence with up to 35 flowering nodes, tightly packed, up to 15 mm wide; bracts woody; calyx erect, not embedded in rachis, calyx 4-toothed plus 1; corolla tube 15 – 20 mm, lobes 4 – 6 mm, blue; anthers at top of tube; fruit surface reticulate, pale, apex extended and flattened with short stylopodium.

This group corresponds to Clade IV. Six species in Brazil.

Group 3 Microphylla. Subshrubs or herbs with or without a woody rootstock; inflorescence with up to 35 flowering nodes, tightly packed, up to 6 mm wide; bracts woody; calyx erect, embedded in rachis, 2-lobed; corolla tube 10 – 14 mm, lobes 5 – 6 mm, red; anthers at top of tube; fruit surface reticulate, pale, apex extended and flattened with short stylopodium.

Two species in Brazil.

Group 4 Quadrangula. Tall shrubs (treelets) to 3.5 m (rootstock unknown); inflorescence with 20 – 26 flowering nodes, tightly packed, narrow (2 – 5 mm wide) to broad (7 – 15 mm wide); bracts herbaceous; calyx erect or held at an angle, 2-lobed; corolla tube 15 – 40 mm, bent and held at right angle to rachis, or straight and held erect; lobes 3 – 5 mm, blue, red or white; anthers at top of tube; fruit surface smooth, pale, beaked at apex, without stylopodium.

This group corresponds to Clade III. Seven species in Brazil.

Group 5 Radlkoferiana. Subshrubs and shrubs (treelet) with woody rootstock; inflorescence with 6 – 12 flowering nodes, tightly packed, broad (10 – 25 mm); bracts herbaceous; calyx not embedded in rachis, 2-lobed; corolla tube 11 – 36 mm, lobes 2 – 4 mm, red, barely spreading; tube straight, erect, with glandular hairs on outer surface; stamens attached below middle of corolla tube; fruit surface reticulate, black, not beaked at apex, without stylopodium; narrow endemics in Bahia.

This group corresponds to part of Clade II. Eight species in Brazil.

Group 6 Villosa. Subshrubs, with woody rootstock; inflorescence with 6 – 35 flowering nodes, tightly packed except in *Stachytarpheta longispicata*, broad (20 – 30 mm), bracts herbaceous; calyx not embedded in rachis, regularly 5-toothed, teeth slightly apiculate, fruiting calyx open; corolla tube 12 – 20 mm, straight; lobes 1 – 4 mm, salmon pink, dark red, dark purple or black, barely spreading; stamens attached half-way up tube; fruit surface reticulate, light or dark brown, with or without short stylopodium, not beaked at apex; mainly from Goiás or Distrito Federal.

Ten species in Brazil.

Group 7 Procumbens. Erect or prostrate herbs from woody rootstock; inflorescence with 6 – 12 flowering nodes, tightly packed, broad, 10 – 15 mm across; bracts herbaceous; calyx 4- or 5-toothed; corolla tube 11 – 30 mm, lobes to 5 mm across, white or very pale blue; style long, far exceeding corolla tube; anthers attached at or about middle of tube; fruit surface reticulate, light or dark brown, without stylopodium.

This group corresponds to Clade VII. Four species in Brazil.

Group 8 Commutata. Shrubs with or without woody rootstock; inflorescence with 6 – 26 flowering nodes, tightly packed, broad, 10 – 20 mm across; bracts woody or herbaceous; calyx 4-toothed, distinctly 4-nerved; corolla tube 10 – 20, lobes 2 – 5 mm; blue; anthers at top of tube; fruit surface reticulate, dark brown, with or without stylopodium.

Eight species in Brazil.

Group 9 Sellowiana. Shrubs with or without woody rootstock; inflorescence with 4 – 6 flowering nodes, barely overtopping upper leaves, tightly packed, broad, 10 – 30 mm wide; bracts woody or herbaceous; calyx 2- or 4-lobed, with sinus on adaxial and abaxial side so that it appears bifid; corolla tube 15 – 20 mm, lobes 3 mm, blue; anthers half-way up tube; fruit surface reticulate, dark brown with stylopodium. From Minas Gerais.

Three species in Brazil.

Group 10 Glabra. Shrubs with or without woody rootstock; inflorescence with 6 – 26 flowering nodes, lax, with calyx held away from the rachis at an acute angle, broad, 10 – 15 mm; bracts woody or herbaceous; calyx 4-toothed abaxially, with outer teeth longer than inner, and sometimes with sinus forming at anthesis between outer and inner teeth; corolla tube 12 – 20 mm, lobes 1 – 4 mm, blue; anthers towards top of tube, just below throat; fruit surface reticulate, brown or black, with stylopodium. From Minas Gerais.

Five species in Brazil.

Group 11 Martiana. Large shrubs, rootstock unknown; inflorescence with 6 – 26 flowering nodes, lax, broad, 10 – 20 mm; bracts woody or herbaceous; calyx with 4 regular teeth on abaxial side; corolla tube 17 – 20 mm, lobes 4 – 6 mm, blue; anthers towards top of tube; fruit surface reticulate, dark brown, with stylopodium and enlarged, prominent commissure. From Minas Gerais, Goiás and Bahia.

The species in this group are similar to those in the Glabra group, but they differ in the calyx, bracts and fruit.

Two species in Brazil.

Group 12 Caatingensis. Shrubs, rootstock not known; inflorescence with up to 26 flowering nodes, rachis narrow, to 1 mm, flowers distant, bracts herbaceous; calyx held patent to rachis, with 4 regular teeth abaxially and narrow sinus adaxially; corolla tube 11 mm, lobes 6 mm, dark purple; anthers at top of tube; fruit surface ± smooth, light brown, apex slightly elongated.

Two species in Brazil.

Taxonomy

Stachytarpheta *Vahl* (1804: 205); Fernandes (1984: 87 – 111), (*nom. conserv.*) over *Valerianoides* Medik. (1789: 177). Type species: *S. jamaicensis* (L.) Vahl based on *Verbena jamaicensis* L. (1753: 19) typ. cons.
Melasanthus Pohl (1828: 75).
Ubochea Baill. (1892: 103).

Woody herbs or shrubs, not aromatic, stems rounded or 4-sided, leaves opposite, simple, crenate or dentate. Inflorescence a terminal spike; flowers sessile, sometimes embedded in the rachis, white, blue, violet, pink, red, black, opening a few at a time, bracteate; bracts conspicuous, often becoming woody and reflexed after anthesis, persistent; calyx tubular, 2, 3, 4 or 5-lobed, equal or unequal, persistent, enveloping fruit; corolla hypocrateriform or infundibular, tube straight or curved, 5-lobed; fertile stamens 2, (posterior), plus 2 staminodes, inserted at middle of tube, thecae divergent. Style persistent, simple; stigma capitate. Ovary bicarpellate, carpel 2-locular, ovules 2, 1 per locule, erect, attached at base; fruit a schizocarp splitting at maturity into 2 hard, narrow 1-seeded mericarps.

Key to groups

1. Inflorescence up to 6 mm wide including calyx and open flower; calyx embedded in excavations in the rachis ··· 2
   Inflorescence broader than 6 mm including calyx and open flower; calyx not, or only partially, embedded in excavations in the rachis ··· 4
2. Calyx with 4 short teeth and 5th tooth or sinus ·············· Group 1. Cayennensis *pro parte* (p. 175)
   Calyx 2-toothed, with a sinus adaxially ··· 3
3. Corollas mauve or blue; fruit apex rounded with short stylopodium ·· Group 1. Cayennensis *pro parte* (p. 175)
   Corollas red; fruit apex extended and flattened with short stylopodium ··· Group 3. Microphylla (p. 208)
4. Inflorescences usually more than 10 cm long ··· 5
   Inflorescences usually less than 10 cm long ··· 12
5. Calyx bifid, i.e. with a sinus on both the adaxial and the abaxial side; fruit pale, smooth, beaked ········
   ·· Group 4. Quadrangula (p. 210)
   Calyx 4- or 5-toothed, sometimes with a sinus on adaxial side; fruit dark, reticulate, not beaked ········ 6
6. Inflorescence with calyx held away from rachis at an acute angle, sometimes up to almost 90° ········· 7
   Inflorescence with calyx held erect, sometimes appresssed to rachis ································ 9
7. Calyx regularly 5-toothed, without a sinus ··················· Group 6. Villosa *pro parte* (p. 225)
   Calyx regularly 4-toothed on abaxial side, and with sinus on adaxial side ······························ 8
8. Rachis narrow (less than 1 mm), with flowers distant (3 – 4 mm apart); corolla tube only slightly longer than calyx; fruit with thin and flat commissure ····················· Group 12. Caatingensis (p. 257)
   Rachis stouter (to 3 mm), flowers not distant (less than 3 mm apart), overlapping; corolla tube at least two thirds longer than calyx; fruit with broad and raised commissure ····························
   ··· Group 11. Martiana *pro parte* (p. 256)
9. Plant decumbent; flowers white ···················· Group 7. Procumbens *pro parte* (p. 237)
   Plant erect; flowers blue ·· 10
10. Inflorescence to 14 cm long; calyx 4-toothed ················ Group 8. Commutata *pro parte* (p. 240)
    Inflorescence to 54 cm long; calyx 5-toothed ··· 11
11. Inflorescence 10 – 15 mm wide; bracts narrowly triangular; fruit with apex extended and flattened ······
    ··· Group 2. Gesnerioides (p. 201)
    Inflorescence 15 – 20 mm wide; bracts broadly ovate; fruit apex not extended and flattened ············
    ·· S. pachystachya (incertae sedis) (p. 261)
12. Calyx 2-lobed ·· 13
    Calyx 4- or 5-lobed ··· 14
13. Flowers bright red; fruit without stylopodium; plants from the Chapadas of Bahia ····················
    ··· Group 5. Radlkoferiana (p. 216)
    Flowers blue; fruit with stylopodium; plants from Minas Gerais ··········· Group 9. Sellowiana (p. 249)
14. Bracts small, inconspicuous; calyx regularly 5-lobed; stamens attached half way up the corolla tube; fruiting calyx not closed at the top ··························· Group 6. Villosa *pro parte* (p. 225)
    Bracts at least half as long as calyx; calyx 4-toothed ; stamens attached at top of the corolla tube; fruiting calyx closed and twisted at the top ··· 15
15. Erect or prostrate herbs from woody rootstock; style long, far exceeding corolla tube; flowers very pale blue or white ··· Group 7. Procumbens *pro parte* (p. 257)
    Low to medium shrubs; style as long as corolla tube; flowers blue ································· 16
16. Bracts 2 – 4 mm long, much shorter than calyx; fruit apex extended and flattened ····················
    ·· S. glandulosa (incertae sedis) (p. 261)
    Bracts 6 – 15 mm long, more than half the length of the calyx; fruit apex rounded ················· 17

17. Inflorescence lax, with the rachis visible between flowers; calyx held at 45° angle to rachis · · · · · · · · · · · ·18
Inflorescence with the flowers tightly packed and obscuring the rachis, calyx held erect · · · · · · · · · · · · ·
· Group 8. Commutata *pro parte* (p. 240)

18. Calyx 4-toothed abaxially, with outer teeth longer than inner, and sometimes with sinus forming at anthesis between outer and inner teeth; corolla tube 12 – 20 mm, lobes 1 – 4 mm; fruit without enlarged commissure · Group 10. Glabra (p. 251)
Calyx 4-toothed abaxially, teeth regular, with no sinus between abaxial teeth; corolla tube 17 – 20 mm, lobes 4 – 6 mm, blue; fruit with enlarged, prominent commissure · · Group 11. Martiana *pro parte* (p. 256)

Group 1 Cayennensis

This forms the biggest group in *Stachytarpheta*, and contains the species which have become naturalised around the world. (This includes the type species of *Stachytarpheta*, *S. jamaicensis*).

Key to species in the Cayennensis Group

1. Corolla tube only slightly longer than calyx · 2
Corolla tube at least twice as long as calyx · 16
2. Calyx at anthesis appearing 2-toothed on abaxial side · 3
Calyx at anthesis with 4 teeth on abaxial side · 4
3. Leaves up to 2 cm wide; leaf margin with 7 – 8 teeth, 2 – 8 mm apart; rachis glabrous; stem hollow · · · · · ·
· 12. **S. angustifolia**
Leaves up to 3 cm wide; leaf margin with 12 – 17 teeth, 2 – 4 mm apart; rachis covered with long, uniseriate hairs; stem not hollow · 13. **S. lythrophylla**
4. Leaves spathulate, linear or linear-elliptic · 5
Leaves oblong or ovate · 7
5. Leaves spathulate, with dense indumentum of uniseriate hairs, especially on undersurface · · 16. **S. spathulata**
Leaves linear or linear-elliptic, ± glabrous · 6
6. Leaves 1.5 – 2 × 0.1 cm; leaf margin inrolled; single nerve on leaf undersurface hardly visible, dark; corolla blue, lobes 2 mm across · 14. **S. linearis**
Leaves 3 – 4 × 0.2 – 0.4 cm; leaf margin not inrolled, single nerve on leaf undersurface prominent, pale; corolla bright purple, lobes 5 mm across · · · · · · · · · · · · · · · · · · · 15. **S. macedoi**
7. Upper and lower leaf surface completely glabrous · 8
Upper and lower leaf surface sparsely strigose to densely matted-hairy, especially on undersurface · · · · 10
8. Leaves oblong, base truncate, sessile and subamplexicaul · · · · · · · · · · · · · · · · · 6. **S. matogrossensis**
Leaves elliptic to ovate, base narrowed, petiolate · 9
9. Leaves elliptic, 1.5 – 4 × 0.5 – 1.2 cm; bracts half as long as calyx at anthesis · · · · · · · · · 11. **S. schottiana**
Leaves ovate, 2 – 5 × 1 – 2.5 cm; bracts equal in length to calyx at anthesis · · · · · · · · · · · 7. **S. laevis**
10. Flowers white · 3. **S. lactea**
Flowers blue, mauve or pink · 11
11. Leaves, stems and rachis fairly densely covered with short, simple hairs; corolla lobes to 5 mm across · · · · ·
· 5. **S. paraguariensis**
Leaves, stems and rachis sparsely strigose, or densely covered with matted long or short uniseriate hairs; corolla lobes 2 – 3 mm across · 12
12. Leaves with upper and lower surface with densely matted hairs · · · · · · · · · · · · · · · 10. **S. hirsutissima**
Leaves with sparse indumentum, or fine covering of long or short hairs, not matted · · · · · · · · · · · · 13
13. Leaves, stems and rachis with sparse indumentum of short hairs; calyx 3 – 6 mm long · · · · · · · · · · · · 14
Leaves, stems and rachis with fine covering of long hairs; calyx 10 mm long · · · · · · · · · · · 9. **S. maximiliani**
14. Herb with woody stem to 3 m; leaves membranaceous to chartaceous; bracts ovate and acuminate; calyx with no adaxial tooth and an adaxial shallow sinus · · · · · · · · · · · · · · · · · · · 4. **S. jamaicensis**
Shrub to 1.5 m; leaves chartaceous; bracts narrowly triangular; calyx with adaxial tooth and no adaxial sinus · 15
15. Leaves narrowly oblong to ovate, 1.3 – 6.5 × 0.5 – 3 cm; inflorescence 2 – 3 mm wide; adaxial tooth of calyx obsolete or reduced · 1. **S. cayennensis**

Leaves broadly ovate, 4 – 10 × 1.8 – 4.5 cm; inflorescence 4 mm wide; calyx distinctly 5-toothed, with
adaxial tooth present · 2. **S. polyura**
16   Upper surface of leaf tuberculate, scabrid, with bulbous-based, short, brown uniseriate hairs · · · · · · · · · · ·
· 17. **S. tuberculata**
Upper surface of leaf smooth, ± glabrous · 17
17.   Plant decumbent · 18. **S. stannardii**
Plant erect · 18
18.   Leaves thick, coriaceous; lower leaf surface with veins forming an intricate network, with fine ciliate
hairs along nerves; bracts elliptic · 19. **S. crassifolia**
Leaves chartaceous; lower leaf surface without intricate network, smooth, glabrous; bracts broadly
triangular · 8. **S. restingensis**

---

1. Stachytarpheta cayennensis (*Rich.*) *Vahl* (1804:
208); Jansen-Jacobs (1988); Dubs (1998); Múlgura de
Romero *et al.* (2003).
Verbena cayennensis Rich. (1792: 105). Type: French
    Guyana, Cayenne, 1792, *Leblond* 356 (lectotype G,
    chosen by Munir, 1992; photograph K).
S. maximiliani var. *ciliaris* Moldenke (1940: 472). Type:
    Brazil, Rio de Janeiro, Jardim Botânico, 4 Dec.
    1928, *L. B. Smith* 1420 (NY).

Sub-shrub to 1.5 m, dichotomously branched. Stem 4-
sided, each face slightly rounded, sparsely white-hairy,
usually with 2 opposite sides more hairy than other 2,
often more hairy at nodes. Leaves chartaceous, narrow-
oblong to oblong to ovate, 1.3 – 6.5 × 0.5 – 3 cm; apex
acute; base attenuate to long attenuate, decurrent into
petiole; margin crenate-serrate, very slightly inrolled;
upper surface glabrous to sparsely hairy with scattered
sessile glands, lower surface with white hairs along all
veins and down into petiole. Inflorescence 14 – 20 (–
25) cm long by 2 – 3 mm wide; bracts woody, narrowly
elliptic to narrowly triangular, 4 – 6 mm long, apex
long-acuminate, margin at base scarious, sometimes
ciliate towards apex. Calyx slightly constricted at throat,
4 – 6 mm long, outer surface glabrous to minutely
hairy, 4-toothed with almost equal teeth, with adaxial
sinus. Corolla blue, pale blue or violet (occasionally
white), with white mark at throat, hypocrateriform,
tube straight, slightly constricted at throat, 5 – 7 mm
long, lobes c. 1 – 2 mm across. Stamens attached at
middle of tube, anthers lying just below throat. Ovary
oblong, about 2 mm long; style about 3 mm long. Fruit
c. 3 mm long, dark brown. Plate 3D.

DISTRIBUTION. Central and South America, from the
Caribbean Islands to Argentina, and widely
naturalised throughout tropical and subtropical
regions of the world (Map 2).
SPECIMENS EXAMINED. ACRE: Rio Moa at Cachoeira
Grande, 27 April 1971, *Prance et al.* 12528 (K).
AMAZONAS: Mun. de Manaus, c. 80 km N de Manaus,
13 July 1992, *Nee* 42951 (K). **BAHIA:** Rod.
Ubaitaba/Maraú, a 24 km de Ubaitaba, 13 Dec. 1967,

da Vinha & Castellanos 21 (K). **GOIÁS:** Rio Verde, 13
Jan. 1968, *Philcox* 3980 (K). Chapada dos Veadeiros, 16
March 1973, *Anderson* 7230 (K). **PARANÁ:** Estação
Marumbi, 2 Jan. 1986, *Cordeiro & Kummrow* 202 (K).
PERNAMBUCO: Mun. de Bonita, 6 March 1996, *Hora &
Campelo* 72 (K). **RIO DE JANEIRO:** Brook trail between
Paineiras and Jardim Botânico, 4 Dec. 1928, *L. B.
Smith* 1420 (K). **RIO GRANDE DO SUL,** Mun. Glorinha,
Rod. BR 101, km 43, 3 March 1997, *Ribas & Pereira*
1838 (K). **SÃO PAULO:** Parque do Estado, Jardim
Botânico do Instituto de Botânica da Secretaria do
Meio Ambiente, 30 Jan. 1968, *Gottsberger* 11-30168 (K).
MEXICO: Chiapas, between Ocosingo and Palenque, 8
April 1980, *Mayo & Madison* 364 (K). **HONDURAS:** along
beaches, 8 Jan. 1931, *Schipp* 679 (K). **GUATEMALA:** Santo
Tomás de Castilla, 26 Feb. 1988, *Marshall et al.* 262

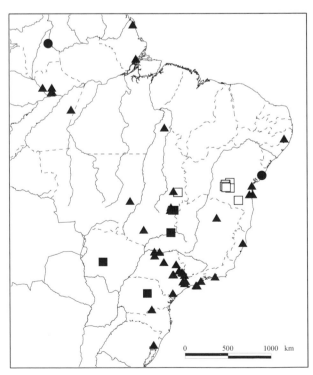

Map 2. Distributions of *Stachytarpheta cayennensis* (▲),
S. jamaicensis (●), *S. lactea* (□) and *S. polyura* (■).

Plate 1. A view from the summit of Pico do Itobira (1930 m), Bahia, with *campo rupestre* in the foreground. **B** *campo rupestre* near Cristalina, Goiás. **C** *Stachytarpheta cassiae*. PHOTOS: R. M. HARLEY.

Plate 2. **A** *Stachytarpheta candida*. **B** *S. macedoi*.   PHOTOS: R. M. HARLEY.

Plate 3. A *Stachytarpheta ganevii*. **B** *S. hispida*. **C** *S. monachinoi*. **D** *S. cayennensis*. PHOTOS: **A** & **C** R. M. HARLEY; **B** S. ATKINS; **D** M. J. E. COODE.

Plate 4. A *Stachytarpheta sericea*. **B** *S. sellowiana*. **C** *S. radlkoferiana*. PHOTOS: **A** & **C** R. M. HARLEY; **B** A. MCROBB.

Plate 5. **A** *Stachytarpheta hatschbachii*. **B** *S. caatingensis*. **C** *S. atriflora*. **D** *S. trispicata*. PHOTOS: R. M. HARLEY.

Plate 6. **A** *Stachytarpheta lychnitis*. **B** *S. galactea*. **C** *S. rupestris*. **D** *S. crassifolia* subsp. *crassifolia*. PHOTOS: **A**, **C** & **D** S. ATKINS; **B** B. L. STANNARD.

Plate 7. **A** *Stachytarpheta longispicata* subsp. *longispicata.* **B** *S. longispicata* subsp. *longispicata.* **C** *S. crassifolia* subsp. *crassifolia.*
PHOTOS: **A** & **B** R. M. HARLEY; **C** S. ATKINS.

Plate 8. A *Stachytarpheta quadrangula*. **B** *S. lacunosa*. **C** *S. coccinea*. PHOTOS: **A** & **C** R. M. HARLEY; **B** S. ATKINS.

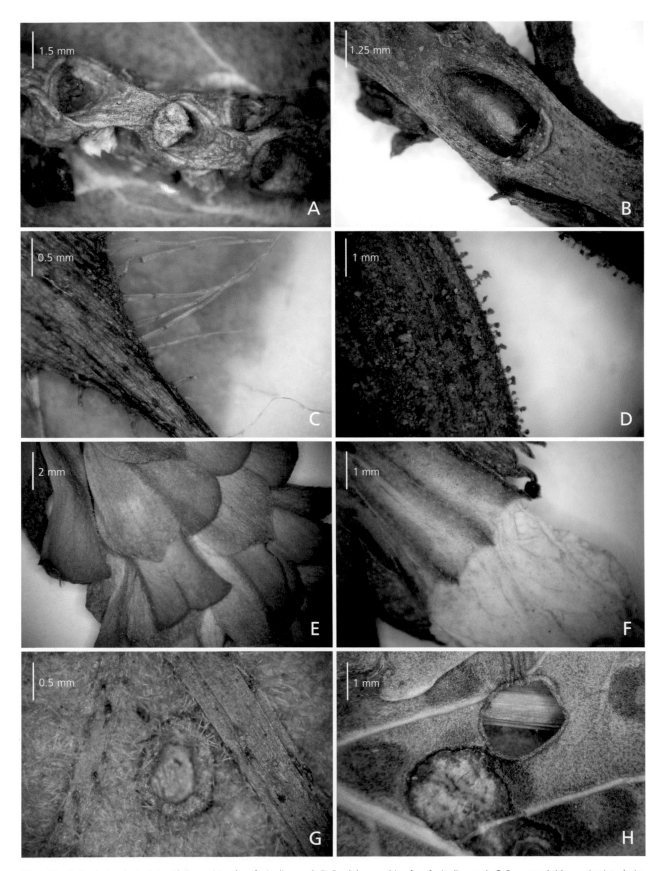

Plate 9. **A** *Stachytarpheta integrifolia*, rachis after fruit dispersal. **B** *S. glabra*, rachis after fruit dispersal. **C** *S. gesnerioides*, uniseriate hairs on bract. **D** *S. sellowiana*, glandular hairs on bract. **E** *S. ganevii*, calyx. **F** *S. confertifolia*, calyx. **G** *S. cearensis*, nectary. **H** *S. atriflora*, nectary.

(K). JAMAICA: Bath, 27 June 1927, *Orcutt* 2003 (K).
TRINIDAD: 13 Feb. 1959, *Richardson* 555 (K). GUYANA:
Mazaruni Station, 14 Aug. 1937, *Sandwith* 1070 (K).
SURINAM: Charlesburg Rift, 8 April 1944, *Maguire &
Stahel* 22771 (K). PARAGUAY: Asunción, 22 Nov. 1974,
Arenas 408 (K). ECUADOR: Napo-Pastaza, Mera, 29 Jan.
1956, *Asplund* 19089 (K). PERU: San Martin, 4 Aug.
1937, *Belshaw* 3197 (K). COLOMBIA: Barranca Bermeja,
26 Feb. 1934, *Haught* 1280 (K). VENEZUELA: Margarita
Island, 6 Aug. 1981, *Sugden* 683 (K). BOLIVIA: La Paz,
29 Sept. 1985, *Solomon & Nee* 14325 (K). ARGENTINA:
Corrientes, 30 Nov. 1987, *Zuloaga et al.* 3317 (K).
HABITAT. Grassland, edge of forest and forest clearings,
roadsides, edge of cultivation and human habitation.
CONSERVATION STATUS. Least concern. The most
widespread and common species to be treated here.
It has become naturalised throughout the tropics and
subtropics.

This species is very variable in size, shape and
indumentum, but the variation does not appear to be
geographically correlated. It has been shown to
hybridise readily (Danser 1929), particularly with
Stachytarpheta jamaicensis.

2. Stachytarpheta polyura *Schauer* (1847: 562; 1851:
201). Type: 'In Brasília merid. in provincia Goyazana,
Meyaponte', *Pohl* s.n. (holotype W).

Subshrub to 1.5 m, dichotomously branched. Stems
rounded; upper part covered with soft white hairs,
especially at nodes, lower part almost glabrous. Leaves
petiolate, often with several younger leaves in axils,
chartaceous, broadly ovate, 4 – 10 × 1.8 – 4.5 cm; apex
obtuse; base long-attenuate, decurrent into petiole;
margin crenate-serrate; upper surface glabrous,
scabrous, lower surface glabrous with scattered hairs
along nerves. Inflorescence long, slender, curving, up
to 30 cm long by 4 mm wide; bracts woody, narrowly
elliptic, 5 mm long, glabrous, margin scarious, ciliate.
Calyx c. 5 – 6 mm long, with scattered white hairs
along nerves, 4 ± equal teeth, and 1 smaller adaxially.
Corolla blue, hypocrateriform, tube straight, c. 5.5 – 7
mm long, slightly constricted at throat, lobes c. 1 mm
across. Stamens almost sessile, attached just below
throat. Ovary ovoid, 2 mm long, style c. 4 mm long.
Fruit dark brown, c. 5 mm long.

DISTRIBUTION. C & S Brazil (Map 2).
SPECIMENS EXAMINED. DISTRITO FEDERAL: Fazenda Água
Limpa (Univ. of Brasília Field Station), near Vargem
Bonita, 24 Sept. 1976, *Ratter et al.* 3657 (K). GOIÁS:
Contraforte Central, 22 Jan. 1970, *Irwin et al.* 25615
(K). MATO GROSSO DO SUL: Mun. Bodoquena, 20 km S
de Bodoquena, 8 Feb. 1998, *Ribas & Pereira* 2579 (K).
PARANÁ: Mun. Jaguariaíva, Recanto Prainha, 10 Feb.

1997, *Ribas & Pereira* 1717 (K).
HABITAT. *Cerrado* and disturbed *cerrado*.
CONSERVATION STATUS. Least concern.

Schauer (1851) stated of this species: "closely related
to *S. cayennensis*, from which it differs in the large
leaves, longer imbricate spikes, longer calyx." There
is a marked difference between these taxa in the size
of the leaves, and in the length and thickness of the
inflorescence spike. The adaxial tooth of the calyx is
always present, whereas in specimens of *S. cayennensis*
the adaxial tooth is often missing, and replaced by a
sinus. I prefer to distinguish this from the more
widespread *S. cayennensis*.

3. Stachytarpheta lactea *Schauer* (1847: 562; 1851:
202); Atkins in Stannard (1995) [as *S. polyura* f.
albiflora]. Type: "in silvis Catingas dictis in Prov.
Bahiensis," *Maximilian of Wied* s.n (holotype M n.v.;
isotype specimen "a" on sheet, BR; photo NY).
Stachytarpheta polyura Schauer f. *albiflora* Moldenke
(1970: 243). Type: Goiás: Chapada dos Veadeiros,
12 Feb. 1966, *Irwin et al.* 12682 (holotype NY;
isotype K).

Shrub to 1.5 m, dichotomously branched. Stems
rounded, lightly covered in long hairs. Leaves
petiolate, patent, chartaceous, oblong, 2.5 – 8 × 0.8 – 3
cm; apex acute to obtuse; base decurrent into petiole;
margin crenate; upper and lower surface sparsely
strigose. Inflorescence up to 23 cm, by c. 5 mm wide,
curving; bracts c. 7 mm long, elliptic, subulate at
apex, glabrous with ciliate margin. Calyx straight, c. 8
mm long with 4 equal teeth and one smaller adaxially,
finely covered with white hairs. Corolla white,
infundibular, tube c. 9 mm long, lobes c. 1 mm across.
Stamens inserted at upper part of tube, ± sessile.
Ovary pyriform. Fruit black, c. 4 mm long.

DISTRIBUTION. Bahia and Goiás (Map 2).
SPECIMENS EXAMINED. BAHIA: Serra do Rio de Contas,
2 km N of Rio de Contas, 27 March 1977, *Harley et al.*
20035 (K); Serra da Conquista, c. 12 km SE of Barra
do Choça, 30 March 1977, *Harley et al.* 20183 (K);
Serra do Sincorá, 6 km N of Cascavel on road to
Mucugê, 25 March 1980, *Harley et al.* 20936 (K); Mun.
Abaíra, Água Limpa, 26 Nov. 1993, *Ganev* 2566 (K);
Mun. Água Quente: Pico das Almas, 22 Dec. 1988,
Harley et al. 27335 (K).
HABITAT. Disturbed woodland, low grassland.
CONSERVATION STATUS. Least concern.

Stachytarpheta cayennensis, *S. polyura* and *S. lactea* are all
close morphologically. *Stachytarpheta lactea* is
differentiated by its milky white flowers, and a
relatively localised distribution.

4. Stachytarpheta jamaicensis *(L.) Vahl* (1804: 206); Jansen-Jacobs (1988).

Verbena jamaicensis L. (1753: 19). Type: Jamaica, (lectotype Herb. Linn. No. 7.13 (S), chosen by Fernandes (1984: 100)).

Herb to 3 m, dichotomously branched. Stem woody, rounded to ± 4-sided, glabrous to sparsely hairy, often more hairy at nodes. Leaves petiolate to subsessile, membranous-chartaceous, ovate, 3 – 10 × 1.8 – 5 cm; apex obtuse; base attenuate, decurrent into petiole; margin crenate; upper and lower surface almost glabrous, with scattered uniseriate hairs (some with ring of glands at base), and sessile glands. Inflorescence curving, up to 50 cm long by 3 – 4 mm wide; bracts woody, elliptic, with long tapering apex, 3 – 5 mm long, glabrous with scarious margin. Calyx 4-toothed, with shallow sinus adaxially, c. 6 mm long, glabrous. Corolla blue, hypocrateriform, tube c. 8 mm long, straight; lobes c. 1.5 mm across. Stamens attached above middle with short filaments. Ovary c. 1 mm long, style c. 6 mm long, slightly exserted. Fruit dark brown to black, reticulate.

DISTRIBUTION. Tropical America (from Tropical Florida to northern South America including the Caribbean Islands) and widely naturalised throughout tropical and subtropical regions of the world. In Brazil (probably naturalised) Bahia, Pernambuco, Roraima (Map 2).
SPECIMENS EXAMINED. BAHIA: Salvador, 2 Dec. 1984, *Guedes & Bromley* 932 (K). **PERNAMBUCO:** 1838, *Gardner* 1105 (E, K). **RORAIMA:** Mun. Bonfim, Nov. 1994, *Milliken* 2285 (K). **U.S.A.:** Florida, Key West, 21 June 1895, *Curtiss* 5424 (K). **GUATEMALA:** Peten, 25 Jan. 1977, *Lundell & Contreras* 20551 (K). **HONDURAS:**Ceiba, 6 July 1938, *Yuncker et al.* 8240 (K). **JAMAICA:** Kingston, Nov. 1849, *Prior* s.n. (K). **CUBA:** Camagüey, 21 July 1996, *Rico et al.* 2034 (K). **GUYANA:** near Georgetown, 1917, *Leechman* s.n. (K); Essequebo river, Oct. 1881, *Jenman* 1103 (K). **SURINAM:** 40 km from Paramaribo, 2 June 1944, *Maguire & Stahel* 23587 (K).
HABITAT. Open areas, grassland, roadside, sea strands and river banks.
CONSERVATION STATUS. Least concern. A common and widespread species.

This species is usually to be found in wetter tropical regions, and one would expect its main distribution in Brazil to be the Amazonian states. However, surprisingly few specimens have been seen from Brazil; it is more common further north in the Guianas and the Caribbean. It is less woody than the preceding species, with larger leaves and a broader rachis.

5. Stachytarpheta paraguariensis *Moldenke* (1948: 473). Dubs (1998). Type: Paraguay, Fuerto Olimpo, 18 Oct. 1946, *Rojas* 13615 (holotype NY; isotype K).

Shrub, 1 – 2 m, dichotomously branched. Stems 4-sided, densely white-hairy on upper stems, lower stems ± glabrous. Leaves subsessile, chartaceous, ovate, 2.5 – 7 × 1.2 – 3 cm; apex obtuse; base long-attenuate, decurrent into petiole; margin crenate; upper and lower surface with indumentum of white hairs, more numerous along veins, and on lower surface. Inflorescence slender, up to 20 cm long by 3 mm wide; bracts woody, narrowly elliptic, 8 mm long, pilose with ciliate margin. Calyx straight, 8 mm long, pilose on outer surface, 5-toothed with 4 ± equal teeth and 1 very small adaxially. Corolla violet-blue, hypocrateriform, tube slightly curved, 9 mm long; lobes c. 5 mm across. Stamens inserted above middle of tube. Ovary narrowly oblong, c. 3 mm long; style 10 mm long. Fruit dark brown, c. 7 mm long.

DISTRIBUTION. Paraguay and southern Brazil (Map 3).
SPECIMENS EXAMINED. MATO GROSSO: Chapada dos Guimarães, 6 Jan. 1996, *Dubs* 2073 (K). **MATO GROSSO DO SUL:** 35 km E of Porto Murtinho, 13 April 1992, *Dubs* 1415 (K). **SANTA CATARINA:** Mun. Gravatal, 15 Oct. 1985, *Pakiornik* 161 (K). **PARAGUAY,** Alto-Paraguay: Chaco, 1906, *Fiebrig* 1482 (K).
HABITAT. *Cerrado* and grassland.
CONSERVATION STATUS. Vulnerable. This species has a scattered distribution in Brazil; there are only three collections. The species occurs in the Chaco region of Paraguay. None of the sites from which it has been collected is specifically protected.

Similar to *Stachytarpheta cayennensis*, but differs in its more hirsute appearance and larger corolla lobes.

6. Stachytarpheta matogrossensis *Moldenke* (1973a: 222). Dubs (1998). Type: Mato Grosso do Sul, Mun. Corumbá, Serra de Urucum, 15 April 1972, *Hatschbach* 29523 (holotype TEX; isotypes MBM, NY).

Herb to 50 cm, unbranched. Stem 4-sided, glabrous. Leaves often with 1 – 3 smaller leaves at a node, patent, sessile, subamplexicaul, chartaceous, oblong, (1.5) 2 – 6 × 0.5 – 1.5 cm; apex acute to obtuse; base truncate; margin entire in lower part of leaf, distantly serrated towards the apex; upper and lower surface glabrous. Inflorescence 14 – 26 cm long by c. 4 mm wide; bracts elliptic, c. 4 mm long, glabrous. Calyx c. 7 mm long, with 4 subulate teeth and sinus adaxially, glabrous. Corolla hypocrateriform, violet with white throat, tube slightly kinked, c. 10 mm long, lobes c. 2 mm across. Stamens inserted at middle of tube, anthers lying just below throat. Ovary c. 1 mm long; style c. 10 mm long. Fruit black with reticulate surface.

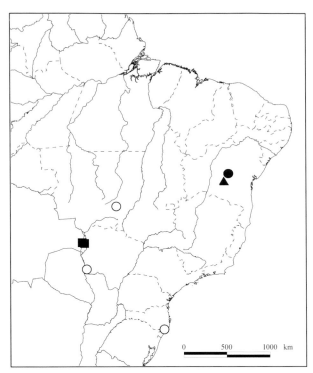

Map 3. Distributions of *Stachytarpheta matogrossensis* (■), *S. paraguariensis* (○), *S. stannardii* (▲) and *S. tuberculata* (●).

DISTRIBUTION. Mato Grosso do Sul (Map 3).

SPECIMENS EXAMINED. MATO GROSSO DO SUL: Mun. Ladário, Fazenda Uruba, 8 June 1994, *Hatschbach et al.* 60787 (MBM, K); Fazenda Bandalta, 24 July 1997, *Damasceno Junior et al.* 1114 (COR, K): Logradouro, Fazenda São Sebastião do Carandá, 11 Oct. 2000, *Damasceno Junior et al.* 2041 (COR, K).

HABITAT. In a rock pavement or in rocky soil.

CONSERVATION STATUS. Near threatened. Not many collections have been made. The surroundings are quite built-up with many roads, towns and an airport.

Superficially very similar to *Stachytarpheta hassleri*, which is endemic to Paraguay. However, the latter is a much bigger plant (1 – 3 m high), with a larger corolla (tube 14 mm long, lobes 5 mm across).

7. Stachytarpheta laevis *Moldenke* (1947b: 369). Type: Rio Grande do Sul, Porto Alegre, 3 Nov. 1892, *Lindman* A 607 (holotype S).

Shrub to 1 m, dichotomously branched. Stem rounded to obscurely 4-sided, glabrous. Leaves somewhat patent, often with further leaves or branchlets in the same axil, chartaceous, ovate, 2 – 5 × 1 – 2.5 cm; apex acute; base attenuate; decurrent into petiole; margin serrate; upper and lower surface glabrous. Inflorescence up to 30 cm long × c. 4 mm wide; bracts linear, c. 7 – 8 mm long, margin scarious, glabrous. Calyx c. 8 mm long, 4

equal teeth, and shallow sinus adaxially, glabrous. Corolla light blue, hypocrateriform, tube straight, c. 9 mm long, lobes c. 3 mm across; anthers lying towards top of tube. Ovary flask-shaped, c. 2 mm long, style c. 7 mm long. Fruit not seen.

DISTRIBUTION. Rio de Janeiro and Rio Grande do Sul (Map 4).

SPECIMENS EXAMINED. RIO DE JANEIRO: Barra da Tijuca, 26 March 1964, *Hoehne* 5648 (NY). **RIO GRANDE DO SUL**: 1897, *Reineck & Czermak* 70 (K); Porto Alegre, 2 Oct. 1948, *Moldenke & Moldenke* 19674 (NY);

HABITAT. In hedgerow thickets and shady roadside places. Sea level.

CONSERVATION STATUS. Vulnerable. Only two localities for this species, separated by some 1500 km, which except for a few National Parks and conservation areas, is one of the busiest coastal regions of the world. If its range once extended along this length of coastline, then it no longer does, and it may well now be a threatened species.

Stachytarpheta laevis, and the next four species are all to be found near the coast in *restinga* vegetation. *S. maximiliani* and *S. hirsutissima* are both hairy, while *S. laevis*, *S. restingensis* and *S. schottiana* are all more or less glabrous. *S. schottiana* can be identified by its much narrower leaves. *S. laevis* and *S. restingensis* are similar in appearance, but can be distinguished by the length of the calyx relative to the corolla tube.

8. Stachytarpheta restingensis *Moldenke* (1959b: 82). Type: Rio de Janeiro, Casimiro de Abreu, near Barra de São João village, 6 Sept. 1953, *Segadas-Vianna et al.* 975 (holotype R n.v.; isotype TEX).

Shrub to 1.5 m, dichotomously branched. Stem 4-sided, glabrous to sparsely hairy on opposite faces. Leaves petiolate, patent, chartaceous, ovate, 4 – 6.5 cm; apex obtuse; base attenuate, decurrent into petiole; upper and lower surface almost glabrous, some specimens with a scattering of fine uniseriate hairs, especially along nerves; margin crenate-serrate, slightly inrolled. Inflorescence 17 – 18 cm long by c. 4 mm wide; bracts broadly triangular, keeled, glabrous with scarious margin. Calyx 4-toothed, ± equal, with shallow sinus adaxially, glabrous, c. 6 mm long. Corolla light blue or mauve, hypocrateriform, tube straight, c. 12 mm long, lobes c. 3 mm across. Ovary flask-shaped c. 1 mm long, style c. 12 mm long. Fruit light brown with reticulate surface, c. 5 mm long. Fig. 4F – K.

DISTRIBUTION. Rio de Janeiro (type) and Espírito Santo (Map 4).

SPECIMENS EXAMINED. ESPÍRITO SANTO: Mun. Conceição da Barra, Itaúnas, 20 May 1999, *Hatschbach et al.* 69210

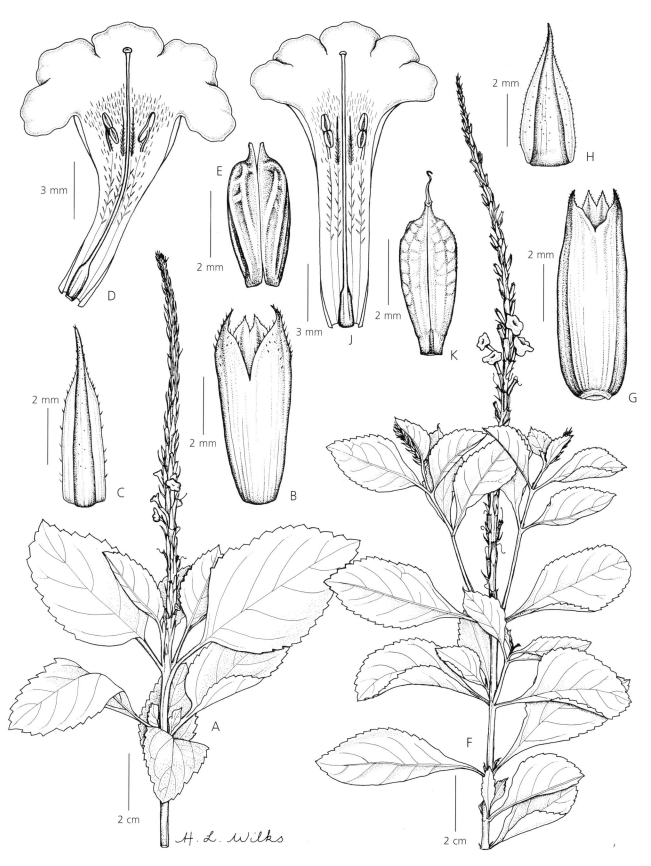

Fig. 4. *Stachytarpheta maximiliani*. **A** habit; **B** calyx; **C** bract; **D** corolla; **E** fruit. *Stachytarpheta restingensis*. **F** habit; **G** calyx; **H** bract; **J** corolla; **K** fruit. **A** from *Hind et al.* 50; **B – E** from *Hatschbach & Motta* 63053; **F – K** from *Hatschbach et al.* 71540. DRAWN BY HAZEL WILKS.

(K); Mun. Itapemirim, Região do Gomes, 23 Oct. 2000, *Hatschbach et al.* 71540 (K); Alfredo Chaves, São Bento de Urânia, 7 Feb. 1999, *Stannard et al.* 1001 (K).

HABITAT. *Restinga* and *restinga arbórea* at sea level.

CONSERVATION STATUS. Endangered. The localities in Espírito Santo are separated by approximately 250 km. In between are tourist beaches, and the wide estuary and dune area of the Rio Doce. This species could be under threat.

NOTE. The stem, leaves, bracts and calyx of *Stannard et al.* 1001 are more hairy than in the type, and this collection may be something new, especially since it is not from a coastal locality.

9. Stachytarpheta maximiliani *Schauer* (1847: 565; 1851: 205). Type: Bahia, "In fruticetis sabulosis etc." 1836, *Blanchet* 2410 (syntype G); 1840, *Blanchet* 3138a (syntype G); 1830, *Salzman* 437 (syntype G).

S. canescens var. *bahiensis* Moldenke (1978: 54). Type: Bahia, Parque Nacional de Monte Pascoal, 26 March 1968, *da Vinha & Santos* 147. (holotype TEX; isotype NY).

S. canescens var. *morii* Moldenke (1979a: 450). Type: Bahia, Santa Cruz de Cabrália, 21 Oct. 1978, *Mori et al.* 10892 (holotype TEX; isotype NY).

S. canescens var. *elliptica* Moldenke (1984a: 233). Type: Bahia, Nova Viçosa, campo de *restinga*, 19 Oct. 1983, *Hatschbach & Guimarães* 47023 (TEX).

S. scaberrima var. *pilosa* Moldenke (1966: 307). Type: Bahia, 32 km W de Canavieiras, 8 Sept. 1965, *Belém* 1757 (holotype TEX).

Shrub to 1.8 m, branching just below inflorescence. Stems woody, rounded, sparsely covered with long, uniseriate hairs. Leaves petiolate, chartaceous, broadly ovate, 2.5 – 7 × 1.7 – 4 cm; apex acute to obtuse; base attenuate decurrent into petiole, margin crenate to serrate; upper surface with sparse indumentum of long uniseriate hairs, lower surface more dense, upper and lower surface with scattered glands. Inflorescence curving, 11 – 21 cm long × c. 4 mm wide; bracts woody, linear, c. 8 mm long, glabrous except for scattered, long hairs along margin. Calyx straight, 10 mm long, 4-toothed with sinus adaxially, outer 2 slightly longer than inner 2, at anthesis teeth sometimes twisted and pressed together, minutely covered with short uniseriate hairs. Corolla mauve, pink or blue, hypocrateriform, tube slightly curved, 10 mm long, lobes c. 2 mm across, glandular on inner surface. Stamens inserted at top of tube, anthers lying just below throat. Ovary flask-shaped c. 2 mm long, style c. 8 mm long. Fruit dark brown, c. 6 mm long. Fig. 4A – E.

DISTRIBUTION. Bahia and Paraíba (Map 4).

SPECIMENS EXAMINED. BAHIA: Reserva Florestal de Porto Seguro, próximo a estrada Mun. km 6950, 4 July

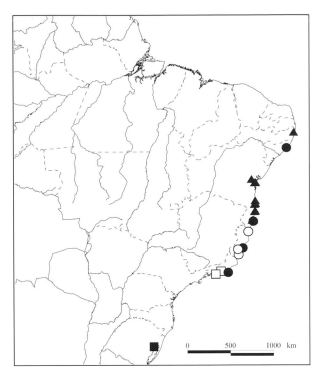

Map 4. Distributions of *Stachytarpheta hirsutissima* (●), *S. laevis* (■), *S. maximiliani* (▲), *S. restingensis* (○) and *S. schottiana* (□).

1990, *Folli* 1193 (K); próximo a estrada Mun. km 1900, 23 April 1991, *Farias* 429 (K); Mun. Valença, Guaibim, 13 Oct. 1998, *Hatschbach et al.* 68489 (K); 13 Oct. 1998, *Hatschbach et al.* 68499 (K); Camino a Guiabim, 4 km de la rodovia Valença – Nazaré, 14 Jan. 1997, *Arbo et al.* 7182 (NY); Mun. de Vera Cruz, Ilha de Itaparica, Praia da Coroa, 31 March 1994, *Melo & Franca,* 937 HUEFS); Canavieiras, 8 Sept. 1965, *Belém* 1757.

PARAÍBA: Mun. Santa Rita, 20 km do centro de João Pessoa, 12 July 1990, *Agra & Gois* 1224 (K).

HABITAT. In clearings and edge of *restinga* forest.

CONSERVATION STATUS. Least concern. Although this is a *restinga* species, and is found along a busy coastal tourist area, there are many collections, and it seems that this plant is able to survive in spite of human encroachment.

This species is sometimes mistaken for *Stachytarpheta jamaicensis*, but the leaves are thicker and more hairy, and the four teeth of the calyx are irregular. The type of *S. canescens* var. *elliptica* from Bahia, Nova Viçosa, is far more hairy than the other specimens seen, but it definitely belongs here.

10. Stachytarpheta hirsutissima *Link* (1821: 19). The holotype in Berlin may have been destroyed, but a specimen at Kew (*Sellow* s.n.) is marked 'Herb. Reg. Berolinense', and has *S. hirsutissima* Link written on it.

S. hirsuta Jacq. f. (1844: 10, pl. 160). Type: Illustration *loc. cit.* Description based on plant grown in Vienna 'Hab. in Brasilia, unde semina retulit H. Schot.'

Shrub or stout herb to 1.5 m, branching just below inflorescence. Stems 4-sided, woody, somewhat gnarled, densely covered with uniseriate hairs. Leaves sessile to short-petiolate, often with smaller leaves in axils, thick-chartaceous, ovate to subrotund, 2 – 4.5 × 1.2 – 3 cm; apex rounded to obtuse; base truncate; margin crenate; upper and lower surface densely matted with white uniseriate hairs. Inflorescence 16 – 30 cm long by 4 – 5 mm wide; bracts woody, narrowly triangular, very hairy with white uniseriate hairs pointing in all directions, patent along margin. Calyx c. 8 mm long with 4 more or less equal teeth and sinus adaxially; densely covered on outer surface with upward pointing uniseriate hairs. Corolla blue, hypocrateriform with tube straight, c. 9 mm long, lobes c. 1.5 mm across. Stamens attached above middle of tube; anthers lying at throat. Ovary c. 1 mm long; style c. 11 mm long. Fruit almost black, c. 7 mm long, outer surface striate and warty. Fig. 5F – K.

DISTRIBUTION. Alagoas, Bahia, Espírito Santo and Rio de Janeiro (Map 4).
SPECIMENS EXAMINED. ALAGOAS: Maceió, Feb. 1838, *Gardner* 1383 (K). **BAHIA:** between Alcobaça and Prado, 15 Jan. 1977, *Harley et al.* 17923 (K). **ESPÍRITO SANTO:** Victora (Vitória), *Sellow* s.n. (K). **RIO DE JANEIRO:** Cabo Frio, Sept. – Oct. 1882, *Glaziou* 13054 (K).
HABITAT. *Restinga.*
CONSERVATION STATUS. Critically endangered. Its range appears to extend from Alagoas in the north down to Rio de Janeiro in the south. Very few collections exist for this species, only one of them modern. It is likely that the pressures on this coastal area have destroyed the habitat in many localities, and this species may be under threat.

This is not *Stachytarpheta canescens* Humb., Bonpl. & Kunth (1815: 227) as suggested by Schauer (1851). *S. canescens* is from Peru, and a different species.

11. Stachytarpheta schottiana *Schauer* (1847: 563; 1851: 202). Type: "In arenosis maritimus prope Rio Jan., 1834, *M. Lund* 104", (syntype G).
S. schottiana var. *angustifolia* Moldenke (1983a: 415). Type: Rio de Janeiro, Mun. Macaé, 5 May 1981, *Araujo* 4415 (holotype NY).
S. restingensis var. *hispidula* Moldenke (1984a: 233). Type: Rio de Janeiro, Maricá, Restinga de Maricá, 21 Jan. 1982, *Landrum* 4170 (holotype NY).

Shrub to 1 m, dichotomously branched. Stem ± 4-sided, woody, glabrous to sparsely hairy with short hairs. Leaves short-petiolate, subcoriaceous, elliptic to narrowly ovate, 1.5 – 4 × 0.5 – 1.2 cm; apex acute to obtuse; base narrowed and decurrent into petiole; margin crenate-serrate; upper and lower surfaces glabrous. Inflorescence 11 – 17 cm long by 3 mm wide; bracts woody, narrowly triangular, c. 4 mm long, glabrous with scarious margin. Calyx 8 mm long, with 4 equal triangular teeth and sinus adaxially, glabrous. Corolla mauve, hypocrateriform, sparsely hairy at throat, tube kinked, 9 mm long, lobes c. 2 mm across. Stamens attached at middle; anthers lying just below throat. Ovary narrowly flask shaped, 2 mm long; style 10 mm long. Fruit dark greyish-black, reticulate, 5 mm long. Fig. 5A – E.

DISTRIBUTION. Rio de Janeiro (Map 4).
SPECIMENS EXAMINED. RIO DE JANEIRO: Recreio dos Bandeirantes, 22 Oct. 1938, *Alston* 146 (NY); 15 Sept. 1948, *Moldenke & Moldenke* 19996 (NY); 15 Sept. 1948, *Moldenke & Moldenke* 19999 (NY); 7 Jan. 1973, *Pabst & Pereira* 9450 (K); Mun. Barra de Tijuca, km 15 W on Rodovia Rio-Santos, 26 Feb. 1988, *Thomas et al.* 6174 (K); Mun. de Macaé, 2 June 1981, *Araujo* 4470 (NY). Without locality 'Rio campos', *Sellow* 223 (K).
HABITAT. A species confined to the coastal *restingas* around Rio de Janeiro.
CONSERVATION STATUS. Critically endangered. This species, apparently confined to the beaches around Rio de Janeiro, is likely to be under threat due to tourist developments.

There is much variation in leaf shape in this species. The type of *Stachytarpheta schottiana* var. *angustifolia* has very narrow leaves, but *Araujo* 4470 from the same locality has slightly broader leaves. *Landrum* 4170, the type of *S. restingensis* var. *hispidula*, has hairs on the stem and leaves.

12. Stachytarpheta angustifolia (*Mill.*) *Vahl* (1804: 205). Schauer (1847: 563; 1851: 203); Jansen-Jacobs (1988). *Verbena angustifolia* Mill. (1768: no. 15). Type: Mexico, La Vera Cruz, *Houston* (BM).
S. elatior Schrad. ex Schult. (1822: 172). Type: In Brasilia, *Prince Maximilian of Wied* s.n. (isotype BR).
S. angustissima Moldenke (1970: 243). Type: Goiás, Serra do Morcego, 20 April 1966, *Irwin et al.* 15104 (holotype TEX; isotype K).

Herb or small shrub to 80 cm, unbranched or branching just below inflorescence. Stem rounded to 4-sided, hollow, glabrous with sparse hairs, and a ring of hairs at each node. Leaves chartaceous, narrow, linear-elliptic, 4 – 9 × 0.5 – 2 cm; apex acute; base attenuate; margin coarsely serrate, with serrations remote; upper surface glabrous with sessile glands, lower surface glabrous with scattered uniseriate hairs

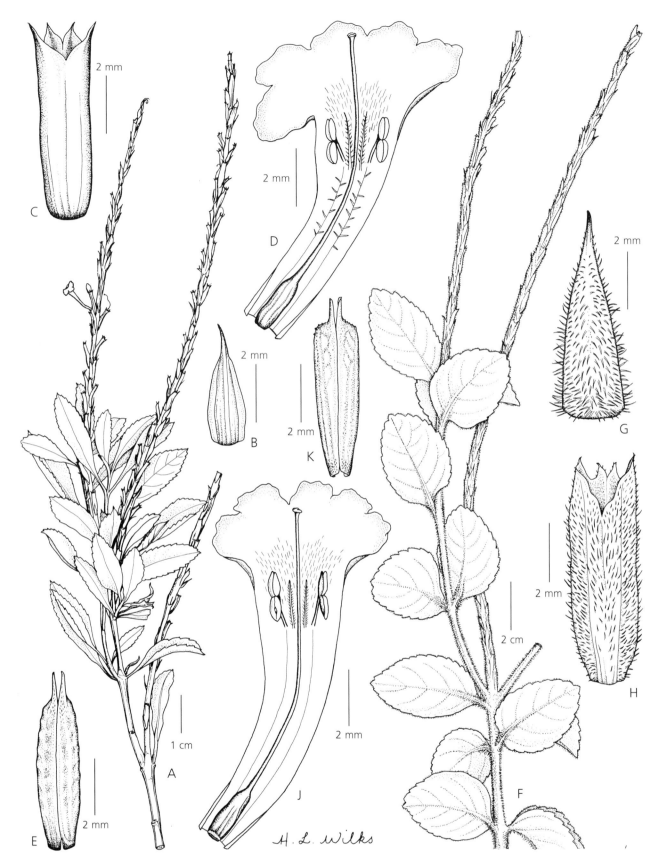

Fig. 5. *Stachytarpheta schottiana*. **A** habit; **B** bract; **C** calyx; **D** corolla; **E** fruit. *Stachytarpheta hirsutissima*. **F** habit; **G** bract; **H** calyx; **J** corolla; **K** fruit. **A** from *Thomas et al.* 6174; **B – E** from *Pabst & Pereira* 9450; **F – K** from *Glaziou* 13054. DRAWN BY HAZEL WILKS.

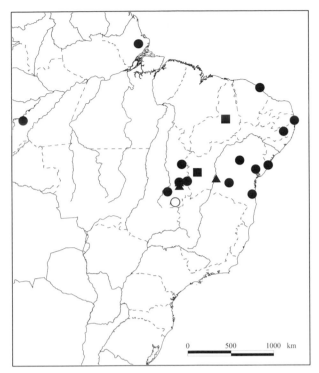

Map 5. Distributions of *Stachytarpheta angustifolia* (●), *S. lythrophylla* (■) and *S. macedoi* (▲).

along nerves. Inflorescence curving, (7 –)14 – 21 cm long by 4 – 6 mm wide, (old inflorescences often still present showing deep excavation scars from fallen flowers), rachis glabrous; bracts woody, narrowly ovate, 3 – 8 mm long, glabrous. Calyx straight, 6 mm long, almost glabrous, appearing 2-toothed with sinus adaxially. Corolla blue, hypocrateriform, tube straight, narrow, c. 10 mm long, glabrous; lobes c. 2 mm wide. Stamens almost sessile, inserted $^3/_4$ way up tube. Ovary oblong, c. 1 mm long; style 10 mm long. Fruit dark brown to black, reticulate, c. 4 mm long.

DISTRIBUTION. Mexico, Caribbean Islands, Central America, C & N South America and West Africa. In Brazil: Alagoas, Amapá, Bahia, Ceará, Goiás, Pará, Paraíba, Pernambuco, Rio de Janeiro (Map 5).
SPECIMENS EXAMINED. 'Brasiliae Tropicae', without date, *Burchell* 9071 (K); Barra, sandy shore, 1857, *Spruce* 1156 (K). **ALAGOAS:** 1838, *Gardner* s.n. (K). **AMAPÁ:** coastal region, 4 Aug. 1962, *Pires & Cavalcante* 52359 (K). **BAHIA:** Mun. de Itabuna, Bairro Jacana, 2 May 1982, *dos Santos*, 12 (K); Oeste de Jacobina, Serra do Tombador, 23 Dec. 1984, *Lewis et al.* CFCR 7477/ SPF 36477 (K); Livramento do Brumado, rodovia para Rio de Contas, 6 April 1992, *Hatschbach & Barbosa* 56659 (K); Feira de Santana, Campus UEFS, beira da Lagoa da Pindoba, 11 May 1995, *França & Melo* 1184 (HUEFS, K). **CEARÁ:** Fortaleza, Lagoa do Tauápe, 12 Aug. 1935, *Drouet* 2218 (K). **GOIÁS:** 1840, *Gardner* 3934 (K); Mun. de Niquelândia, 17 Sept. 1996, *Aparecida da*

Silva et al. 3093 (K). **PARÁ:** Monte Alegre, 9 May 1953, *Lima* 53-1472 (K). **PARAÍBA:** João Pessoa, Mangabeira, 16 July 1991, *de Moura* 639 (K). **PERNAMBUCO:** Oct. 1837, *Gardner* 1106 (K); Mun. de Caruaru, Brejos dos Cavalos, 5 Sept. 1995, *Andrade & Andrade* 210 (K). **RIO DE JANEIRO:** 1882, *Glaziou* 13062 (K). **CUBA:** Santa Clara, 13 Sept. 1895, *Combs* 589 (K). **VENEZUELA:** Cerro Pichacho, 31 Jan. 1961, *Steyermark* 88905 (K).
HABITAT. In damp ground and marshy places, usually around river banks at sea level and lower altitudes.
CONSERVATION STATUS. Least concern. A fairly common species with a wide distribution around tropical America.

Stachytarpheta angustifolia (Mill.) Vahl seems to be the earliest name for this taxon. First described by Miller (1768) as *Verbena angustifolia* "with naked, fleshy spikes and narrow spear-shaped leaves with worn-out sawed edges". This was transferred to *Stachytarpheta* by Vahl (1804). He included in the synonymy *Verbena indica* of Jacquin (1771), but not *Verbena indica* of Linnaeus (1759, 1762). Linnaeus (1762) described *Verbena indica* as "Simillima *V. jamaicensi*, sed caulis laevis abseque pilis. Folia glabra nec retrorsum scabra, magis lineata, non serrata, sed dentibus obtusis obliquis dentata; basi sensim angustata, nec petiolis late marginatis." Vahl described *S. angustifolia* as "foliis lanceolatis utrinque attenuatis remote serratis glabris. Habitat in America meridionalis." He described *S. indica* as "foliis lanceolato-oblongis basi attenuatis remote dentatis cauleque glaberrimus, bracteis lineari-lanceolatis. Habitat in Zeylona et Guinea." (Zeylona [Ceylon] and Guinea refers to west Africa).

Fernandes (1984) presented good reasons for choosing LINN 35-1 as the type for *Stachytarpheta indica* (it has the determination of Linnaeus himself on the sheet.) However, LINN 35-1 does not match Linnaeus' own description. The specimen is definitely the taxon found in South America, with narrow leaves with teeth remote and serrate, (not with oblique, obtuse teeth.)

There are two distinct taxa involved. The first is described here; the second, which seems to occur only in Africa, has oblong to ovate leaves, with crenate margin with obtuse teeth. Both species appear to be herbaceous annuals, with hollow stems, a deeply excavated rachis, inhabiting riversides and seasonally flooded pasture. *Stachytarpheta angustifolia* is a common plant across Brazil and is also found in West Africa. *Stachytarpheta indica* (*sensu* Linnaeus) is also common in West Africa and East Africa, but I have found no corresponding specimens in Brazil. It is possible that *S. indica* was a very early introduction of *S. angustifolia* (they share the same calyx morphology), from South America, and has developed a different morphology. It may also be a hybrid between *S. angustifolia* and *S. jamaicensis*.

There is a sericeous form which Moldenke called *Stachytarpheta angustissima*, but only one collection of this form exists.

13. Stachytarpheta lythrophylla *Schauer* (1847: 563: 1851: 204). Type: 'In fields and pastures of Equatorial Brazil, on river Itapicura, prov. Maranha and upper Prov. Piauí.', *Martius* s.n. (holotype M n.v.; photo. NY).

Annual herb to 60 cm, unbranched or branching just below inflorescence. Stems pale, 4-sided, with scattered uniseriate hairs, more numerous at nodes. Leaves sessile, patent, often with smaller leaves in the axils of larger leaves, elliptic, (2 –)5 – 8 × (0.5 –)1.3 – 3 cm; apex acute; base attenuate, decurrent into petiole; margin closely serrate; upper and lower surface almost glabrous, some younger leaves with scattered uniseriate hairs, especially along veins. Inflorescence 10 – 19 cm long by 4 mm wide, rachis covered with long, uniseriate hairs; bracts subulate, 7 – 8 mm long, glabrous. Calyx 9 mm long, appearing 2-toothed. Corolla blue, hypocrateriform, tube c. 9 mm long, lobes 1.5 mm across. Stamens attached at top of tube, just below throat. Ovary oblong, c. 1 mm long. Fruit pale yellow-brown, 5 mm long. Mature fruit deeply embedded in excavations in the rachis.

DISTRIBUTION. Piauí and Bahia (Map 5).
SPECIMENS EXAMINED. PIAUÍ: Oeiras, 1839, *Gardner* 2285 (K); *Gardner* 2284 (G). BAHIA: Barreiras, 10 km North, 12 March 1979, *Hatschbach* 42090 (NY).
HABITAT. In level plains and fields, close to river.
CONSERVATION STATUS. Vulnerable. The two localities are separated by about 1000 km. The intervening area is fed by many rivers, and could provide the right habitat, but has not been collected because of problems of access. At the moment, the paucity of specimens suggests that this species may be under threat.

This is similar to the more widespread *Stachytarpheta angustifolia*, but differs in the larger leaves with closely serrate margin, and by the hairs on the rachis.

14. Stachytarpheta linearis *Moldenke* (1972: 454). Type: Minas Gerais, Mun. Gouveia, 21 Jan. 1972, *Hatschbach, Smith & Ayensu* 29098 (holotype TEX; isotype MBM).

Small shrub to 30 cm, with xylopodium, branching from base. Stem woody, 4-sided, glabrous with light covering of dark sessile glands. Leaves sessile, erect, somewhat fleshy, linear, 1.5 – 2 × 0.1 cm; apex acute; base truncate; margin inrolled; upper and lower surface glabrous with scattered dark, sessile glands. Inflorescence 7 – 8 cm long by 4 mm wide, bracts narrowly triangular, c. 5 mm long, glabrous. Calyx 8

mm long, 5-toothed, 4 ± equal, 1 smaller adaxially, outer surface glabrous. Corolla blue, hypocrateriform, tube slightly curved, c. 9 mm long, lobes c. 2 mm across. Stamens with anthers lying just below throat. Ovary flask-shaped c. 2 mm long, style c. 8 mm long. Fruit light brown, 5 mm long.

DISTRIBUTION. Minas Gerais.
SPECIMENS EXAMINED: MINAS GERAIS: Mun. Diamantina, Rod. Guinda – C. Mata, 17 Feb. 1973, *Hatschbach & Ahumada* 31700 (K, NY); Estrada para Conselheiro Mata, 6 km da estrada Diamantina – Curvelo, 18 Nov. 1984, *Stannard et al.* CFCR 6122/SPF 35722 (K).
HABITAT. In seasonally flooded grassland near river.
CONSERVATION STATUS. Endangered. A very small area of occurrence, and very few individuals in each locality. Grazed by cattle.

15. Stachytarpheta macedoi *Moldenke* (1950a: 276). Type: Minas Gerais, Mun. Ituiutaba, São Terezinha, 11 Feb. 1949, *Amado Macedo* 1647 (holotype NY).

Slender erect herb to 20 cm, unbranched. Stems herbaceous, square, glabrous. Leaves sessile, patent, decussate, chartaceous, narrowly elliptic, 3 – 4 × 0.2 – 0.4 cm; apex acute; base cuneate to truncate; margin entire; upper and lower surface glabrous. Inflorescence 8 – 12 cm long × 3 mm wide; bracts linear, 5 mm long, glabrous. Calyx straight, 7 mm long, 4-toothed with sinus adaxially, glabrous. Corolla bright purple, infundibular, tube curving, c. 6 mm long, glandular at throat, lobes c. 5 mm across. Stamens attached above middle. Ovary elongated, 1.5 mm long. Fruit dark brown. Plate 2B.

DISTRIBUTION. Bahia, Goiás, Minas Gerais and Tocantins (Map 5).
SPECIMENS EXAMINED: BAHIA: Basin of Upper São Francisco River, 32 km NE from Bom Jesus de Lapa, 18 April 1980, *Harley et al.* 21477a (K). GOIÁS: N of Funil and the Rio Parana, 14 March 1973, *Anderson et al.* 7109 (K). TOCANTINS: Rodovia Arraias – Paraná, 12 Feb. 1994, *Hatschbach & Silva* 60512 (CTES).
HABITAT. In seasonally flooded, grazed open fields, on sandy soil.
CONSERVATION STATUS. Endangered through spread of cultivation.

This is a pretty herb which often grows alongside *Stachytarpheta angustifolia*, to which it is similar, but is a more delicate plant.

16. Stachytarpheta spathulata *Moldenke* (1964: 76). Type: Brazil, Minas Gerais, 18 km SW of Diamantina, 10 April 1973, *Anderson* 8515 (holotype TEX; isotype NY).

Shrub to 1.2 m, branched. Stems woody, rounded, covered in short, white uniseriate hairs. Leaves short-petiolate, thick-chartaceous, spathulate to fan-shaped, 7 – 20 × 6 – 16 mm; apex rounded; base attenuate or truncate-attenuate; margin crenate, slightly inrolled; upper surface covered with dense indumentum of uniseriate hairs, lower surface more dense. Inflorescence 1 – 10 cm long × 4 – 5 mm wide, bracts woody, ovate or narrowly triangular, 4 mm long, with uniseriate hairs on surface. Calyx 8 – 12 mm long, with 4 teeth and sinus adaxially, outer surface covered with uniseriate hairs and glands. Corolla blue, hypocrateriform, tube c. 15 mm long, slightly enlarged at middle, lobes c. 2 mm across. Stamens inserted at top of tube. Ovary oblong, c. 2 mm long; style long c. 16 mm long. Fruit dark brown, surface slightly reticulate.

Shrub to 1 m; leaves spathulate, 7 – 12 × 7 – 16 mm; inflorescence 4 – 10 cm long by 4 mm wide · subsp. **spathulata**
Shrub to 1.2 m; leaves fan-shaped, 8 – 20 × 6 – 15 mm; inflorescence 1 – 6 cm long by 4 – 5 mm wide · subsp. **mogolensis**

a. subsp. **spathulata**

Shrub to 1 m. Leaves short-petiolate, thick-chartaceous, spathulate, 7 – 12 × 7 – 16 mm, apex rounded, base truncate-attenuate. Inflorescence 4 – 10 cm long by 4 mm wide, bracts woody, narrowly triangular; calyx 12 mm long.

DISTRIBUTION. Minas Gerais.
SPECIMENS EXAMINED. MINAS GERAIS: 1816 – 1821, *Saint-Hilaire*, Catal. B1, No. 2041 (K); Mun. Diamantina, estrada para Conselheiro Mata, 19 Nov. 1984, *Stannard et al.* CFCR 6178 (K); Rodoviario Guinda – Conselheiro Mata, 19 March 1997, *Hatschbach et al.* 66476 (K).
HABITAT. *Campo rupestre* (quartzitic).
CONSERVATION STATUS. Endangered. This species has a small area of occurrence, and should be protected. This little valley has other endemic species (see also *Stachytarpheta linearis* and *S. itambensis*), is close to Diamantina, and is grazed by animals.

A very attractive shrub with bluish-green foliage and dark blue flowers.

b. subsp. **mogolensis** *S. Atkins* **subsp. nov.** a subsp. *spathulata* longitudine inflorescentiae 1 – 6 cm (non 4 – 10 cm), foliis flabellatis non spathulatis differt. Typus: Minas Gerais, Grão Mogol, a sudoeste da cidade, 21 May 1982, *Mamede et al.* SPF 23593/CFCR 3388 (holotypus SPF; isotypus K).

Shrub to 1.2 m, branched. Leaves petiolate, thick chartaceous, fan-shaped, 0.8 – 2 × 0.6 – 1.5 cm; apex rounded; base attenuate. Inflorescence 1 – 6 cm long by 4 – 5 mm wide, bracts ovate, c. 4 mm long. Calyx c. 8 mm long.

DISTRIBUTION. Minas Gerais.
SPECIMENS EXAMINED. MINAS GERAIS, Grão Mogol, 12 April 1981, *Furlan et al.* CFCR 751/SPF 22685 (K) & 15 April 1981, *Cordeiro et al.* CFCR 937/SPF 22864 (K) & 15 Oct. 1988, *Harley et al.* 25060 (K); Mun. Cristália, Estrada Grão Mogol – Cristália, 18 July 1998, *Hatschbach et al.* 67985 (K); 14 Feb. 2003, *França et al.* 4347 (HUEFS; K); Botumirim, saída ao sul da cidade, 19 Nov.1992, *Mello-Silva et al.* 694 (K).
CONSERVATION STATUS. Endangered. Occurs only in a small area of *campo rupestre*, on the road between Grão Mogol and Cristália (a distance of about 25 km).

This subspecies has consistently shorter, less conspicuous inflorescences than the typical subspecies.

17. Stachytarpheta tuberculata *S. Atkins* **sp. nov.** a *S. crassifolia* inflorescentiis brevioribus usque 4 cm (non usque 25 cm), pagina superiore foliorum juvenilium pilis basi bulbosis obsita (non glabra), rachidi inflorescentiae indumentosa (non glabra) distincta. Typus: Bahia, Mun. Palmeiras, Morro do Pai Inácio, 9 July 1996, *Hind et al.* PCD 3518-b (holotypus HUEFS; isotypus K).

Branched shrub to 1.3 m. Stems 4-sided, "gnarled", striated, enlarged at nodes, upper part covered in hairs with dark bulbous bases, lower part glabrous. Leaves sub-coriaceous, ovate, 2 – 4 × 1 – 2 cm; apex obtuse to rounded; base narrow-cuneate; upper surface tuberculate with hair perched on top of each tubercule, lower surface with network of raised veins interspersed with hairs, some white, some with dark bases; margin serrate, slightly inrolled. Inflorescence c. 4 cm long × 4 mm wide; bracts woody, narrow, c. 6 mm long, hairy with single dark line at keel. Calyx c. 10 mm long, slightly woody, hairy on outer surface, 4-toothed with outer 2 longer. Corolla pink-blue, hypocrateriform, tube straight, c. 20 mm long, lobe c. 2 mm across, hairy at throat. Stamens attached just below throat. Ovary flask-shaped, c. 2 mm long. Fruit light brown, reticulate. Fig. 6.

DISTRIBUTION. Bahia: Mun. Palmeiras, Pai Inácio (Map 3).
SPECIMENS EXAMINED. BAHIA: Mun. Palmeiras, Morro do Pai Inácio, June 1986, *Santos* s.n. (HUEFS); Pai Inácio, 21 Nov. 1994, *Melo et al.* PCD 1196 (K); without date, *Alves et al.* 4238 (K).

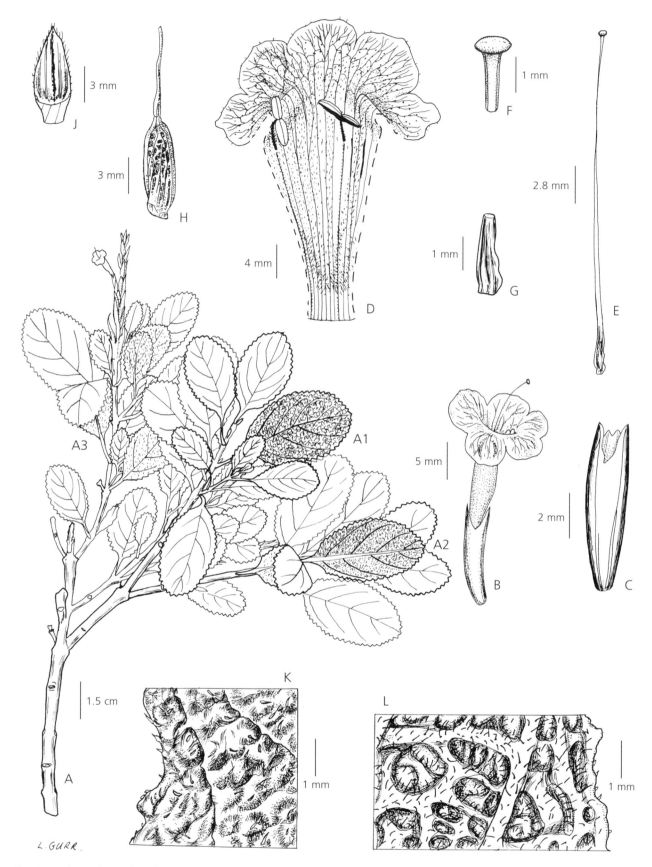

Fig. 6. *Stachytarpheta tuberculata*. **A** habit; **B** calyx & corolla; **C** calyx; **D** corolla; **E** ovary, style & stigma; **F** apex of style; **G** ovary; **H** fruit; **J** bract; **K** detail of upper leaf surface; **L** detail of lower leaf surface. From *Melo et al.* PCD 1196. DRAWN BY LINDA GURR.

HABITAT. *Campo rupestre*, among rocks at 1000 – 1800 m.
CONSERVATION STATUS. Vulnerable. Conservation Dependent. Pai Inácio is not inside the National Park, but is an APA (Área de proteção Ambiental), which means that the local people can use the land in a "sustainable" way. Pai Inácio is the property of Igreja Universal do Reino de Deus, and access to the top is restricted.

A species similar to *Stachytarpheta crassifolia*, but with a short inflorescence (to 4 cm) and with brown hairs, which make it look as if the plant might be viscid. Very restricted locality.

18. Stachytarpheta stannardii S. *Atkins* **sp. nov.** a *S. crassifolia* habitu fruticis decumbentis (non erecto), longitudine calycis 10 mm (non 6 – 8 mm), tubo corollae 20 mm longo (non 12 – 16 mm) distinguenda. Typus, Bahia, Mun. Abaíra, Riacho da Taquara, 28 Jan. 1992, *Stannard et al.* H 50835 (holotypus SPF; isotypi CEPEC, HUEFS, K).

Decumbent or prostrate shrub to 40 cm. Stems 4-sided, woody, often shedding outer layer, fairly densely covered with short uniseriate hairs. Leaves short-petiolate, coriaceous, ovate, 2 – 3(– 5) × 1.5 – 2(– 3.5) cm; apex obtuse to rounded, occasionally emarginate; base truncate or angustate, decurrent into petiole; margin crenate; upper surface glabrous, lower surface with intricate network of secondary veins, densely hairy in between, hairs uniseriate with dark bases. Inflorescence 15 – 30 cm long × 5 mm wide; bracts narrowly triangular, c. 5 mm long, with scattered short hairs, margin ciliate. Calyx c. 10 mm long, with 4 equal teeth abaxially and sinus adaxially, outer surface sparsely strigose and glandular. Corolla tube purple, lobes dark blue, hypocrateriform, tube c. 20 mm long, lobes 4 mm across. Stamens attached at middle of corolla tube, anthers lying just below throat. Ovary flask-shaped, 2 mm long; style 18 mm long. Fruit light brown, c. 6 mm long with prominent attachment scar. Fig. 7.

DISTRIBUTION. Bahia: Mun. Abaíra (Map 3).
SPECIMENS EXAMINED. BAHIA: Mun. Abaíra, Tijuquinho, 4 March 1992, *Sano & Læssøe* H50861 (K); Descida para Tijuquinho, 6 Jan. 1992, *Harley et al.* H 50650 (K); Serra ao Sul do Riacho da Taquara, 10 Jan. 1992, *Harley et al.* H 51242 (K).
HABITAT. *Campo rupestre*, among rocks at 1700 – 1900 m.
CONSERVATION STATUS. Endangered. Only found in two localities, some 10 km apart; could be under threat from encroaching agriculture.

Another species similar to *Stachytarpheta crassifolia*, but a decumbent shrub with a longer calyx and corolla.

This species is named for my colleague Brian Stannard who collected the type specimen.

19. Stachytarpheta crassifolia *Schrad.* (1821: 709). Schauer (1847: 566; 1851: 206); Zappi *et al.* (2003). Schrader did not specify a type, but worked on material brought back from Brazil by Prince Maximilian of Wied. The specimen Schrader worked on is probably in LE (n.v.). Schauer cites *Blanchet* 3647 (G, K).

Branched shrub to 2 m. Stems woody, rounded to 4-sided, glabrescent. Leaves petiolate, erect to patent, coriaceous, elliptic to narrowly ovate to broadly ovate to rotund, to obovate, 1.5 – 7.5 × 1 – 8 cm, apex rounded or obtuse, base attenuate into petiole, margin serrulate, slightly inrolled, upper surface glossy, glabrous, lower surface with white hairs along nerves, secondary veins forming an intricate reticulate network. Inflorescence 9 – 20(– 30) cm long by 4 – 5 mm wide; bracts woody,

Table 3. Showing differences between *Stachytarpheta crassifolia* subsp. *crassifolia*, subsp. *abairensis*, subsp. *rotundifolia*, and subsp. *minasensis*

Taxon	Leaf shape	Leaf size (cm)	Infl. width (mm)	Calyx length (mm)	Corolla tube length (mm)	Geography	Plant size (m)	Altitude (m)
S. crassifolia subsp. *crassifolia*	narrow-ovate, base long attenuate	3 – 9 × 1.5 – 3.5	4 – 5	6 – 8	12 – 16	Bahia: Mun. Jacobina, Lençois & Mucugê	0.5 – 2	550 – 1000
S. crassifolia subsp. *abairensis*	ovate to broadly ovate, base cuneate	4 – 9 × 2.6 – 4.5	5 – 6	9	16	Bahia: Mun. Abaíra & Rio de Contas	1.5 – 2	1250 – 1850
S. crassifolia subsp. *rotundifolia*	broadly ovate to rotund, base cuneate	2 – 4 × 1.5 – 3	3 – 4	9	20	Bahia: Mun. Morro do Chapeu	1 – 1.5	900 – 1100
S. crassifolia subsp. *minasensis*	ovate, base attenuate	4.5 – 6.5 × 2 – 2.8	5 – 6	9	20	Minas Gerais: Mun. Jequitinhonha	0.5 – 2	1100

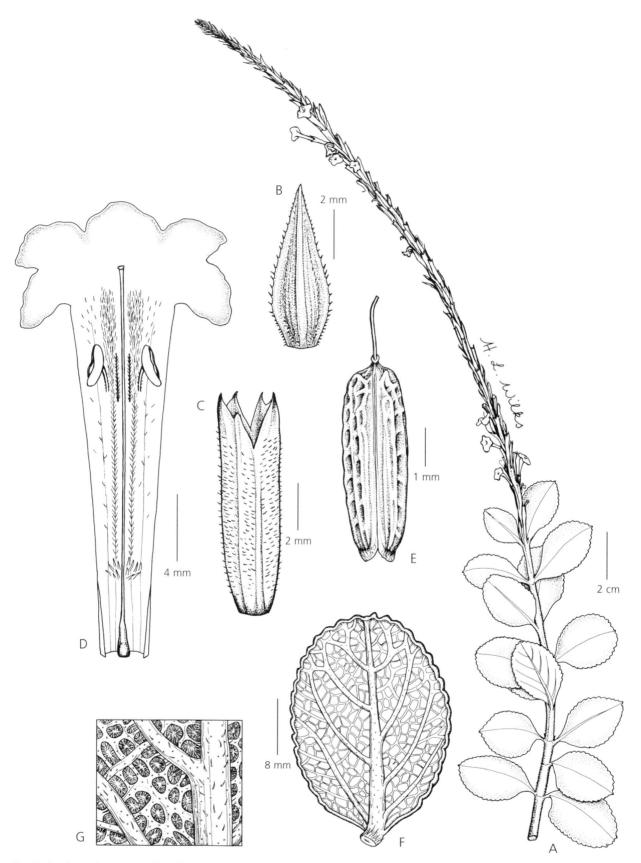

Fig. 7. *Stachytarpheta stannardii*. **A** habit; **B** bract; **C** calyx; **D** corolla; **E** fruit; **F** leaf; **G** detail of leaf undersurface. **A**, **F** & **G** from *Harley et al.* H51242; **B** – **E** from *Sano & Læssøe* **H** 50861. DRAWN BY HAZEL WILKS.

elliptic, acuminate, c. 5 mm long. Calyx c. 6 – 8 mm long, 4-toothed, ± equal, outer surface covered with short white hairs interspersed with sessile glands. Corolla intense blue or white, hypocrateriform, tube straight, c. 12 – 20 mm long, lobes 5 ± equal c. 2 – 5 mm across. Stamens inserted at throat. Ovary c. 2 mm long, style c. 13 mm long, included. Fruit black, surface reticulate, attachment scar not prominent.

This species varies in leaf shape and size, calyx and corolla length and inflorescence width. When viewed all together, it is impossible to include all the specimens in a single taxon. The differences can be correlated with geography, and so I have decided to recognise the differences at subspecific rank. See Table 3.

a. subsp. **crassifolia**

Branched shrub to 2 m. Stems woody, rounded to 4-sided, glabrescent. Leaves petiolate, erect to patent, coriaceous, elliptic to narrowly ovate-obovate, 2 – 6(– 7.5) × 1 – 2(– 3.5) cm, apex rounded or obtuse, base attenuate into petiole, margin serrulate, slightly inrolled, upper surface glossy, glabrous, lower surface with white hairs along nerves, secondary veins forming an intricate reticulate network. Inflorescence 9 – 20(– 25) cm long × 4 – 5 mm wide; bracts woody, elliptic, acuminate, c. 5 mm long. Calyx c. 6 – 8 mm long, 4-toothed, ± equal, outer surface covered with

Map 6. Distributions of *Stachytarpheta crassifolia* subsp. *crassifolia* (●), *S. crassifolia* subsp. *abairensis* (▲), *S. crassifolia* subsp. *minasensis* (■) and *S. crassifolia* subsp. *rotundifolia* (○).

short white hairs interspersed with sessile glands. Corolla intense blue, hypocrateriform, tube straight, c. 12 – 16 mm long, lobes 5 ± equal c. 2 mm across. Stamens inserted at throat. Ovary c. 2 mm long, style c. 13 mm long, included. Fruit black, surface reticulate, attachment scar not prominent. Fig. 8A, N. Plates 6D, 7C.

DISTRIBUTION. Bahia (Map 6).
SPECIMENS EXAMINED. BAHIA: "ad Jacobina et Moritiba "*Blanchet* 3647 (K; BR); Mun. Jacobina, Hotel Serra de Ouro, 23 Oct. 1990, *Freire-Fierro et al.* 2064 (K); Mun. Lençóis: Estrada Lençóis – Fazenda Remanso, 28 Oct. 1978, *Martinelli et al.* 5362 (K); 4 km NE of Lençóis, 23 May 1980, *Harley et al.* 22456 (K); 7 – 10 km along Seabra-Itaberaba road, 27 May 1980, *Harley et al.* 22667 (K); on trail to Barro Branco, 5 km N of Lençóis, 13 June 1981, *Mori & Boom* 14389 (K); 1 km leste da estrada Lençóis BR 242, 5 July 1983, *Coradin et al.* 6499 (K); Estrada entre Lençóis e Seabra, 14 Feb. 1994, *Atkins et al.* CFCR 14055 (K); Morro da Chapadinha, 22 Nov. 1994, *Melo et al.* 1248 (K); 31 Aug. 1994, *Guedes et al.* 699. Mun. de Mucugê, by Rio Cumbuca S of Mucugê, 4 Feb. 1974, *Harley et al.* 15896 (K); S of Andaraí on road to Mucugê, 14 Feb. 1977, *Harley et al.* 18668 (K); estrada de Mucugê para cascavel, 20 March 1990, *de Carvalho & Saunders* 2954 (K); 3 – 5 km N of Mucugê em direção a Palmeiras, 20 Feb. 1994, *Atkins et al.* CFCR 14273 (K).
HABITAT. *Campo rupestre*, often on sand, and close to rivers, at altitudes up to 1000 m.
CONSERVATION STATUS. Least concern. A common species around Lençóis and through the Parque Nacional da Chapada Diamantina.

b. subsp. **abairensis** *S. Atkins* **subsp. nov.** a subsp. *crassifolia* inflorescentiis crassioribus (4 – 5 mm crassis non 2 – 3 mm), floribus majoribus tubo corollae usque 20 mm longo (non 15 mm) lobis c. 5 mm latis (non 2 mm) differt. *S. crassifolia sensu* S. Atkins in Stannard (1995: 628). Typus: Bahia: Pico das Almas, 19 March 1977, *Harley et al.* 19683 (holotypus SPF; isotypi CEPEC; K).

Shrub to 2 m. Stem stout, 4-sided. Leaves thick, ovate to broadly ovate, 3 – 7.5 × 3.2 – 8 cm. Inflorescence up to 30 cm long × 4 – 5 mm wide; corolla tube up to 2 cm, lobes c. 5 mm across. Fig. 8M.

DISTRIBUTION. Bahia: Abaíra, Rio de Contas (Map 6).
SPECIMENS EXAMINED. BAHIA. Mun. Abaíra, Guarda Mor, 15 Sept. 1993, *Ganev* 2249 (K); Caminho Boa Vista para Bicota, 9 July 1995, *França et al.* 1273 (HUEFS; K); Serra dos Cristais, 8 June 1994, *Ganev* 3336 (K); Água Limpa, 21 Dec. 1991, *Harley et al.* H 50246 (K); Serra em Catolés de Cima, 17 April 1994, *França et al.* 995 (HUEFS; K); Jambreiro, 31 March 1994, *Ganev* 3015

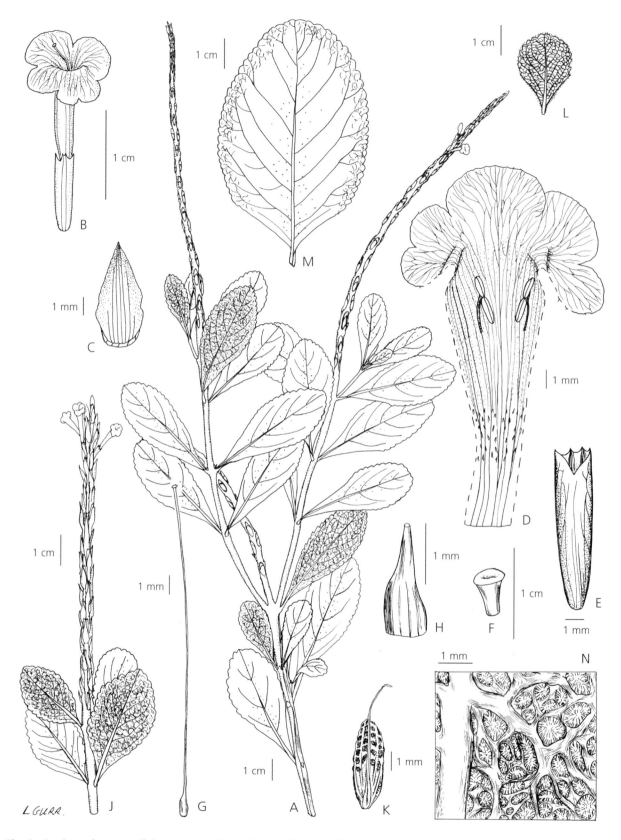

Fig. 8. *Stachytarpheta crassifolia* subsp *crassifolia*. **A** habit; **N** detail of lower leaf surface. *S. crassifolia* subsp. *minasensis*. **B** flower; **C** bract; **D** corolla; **E** calyx; **F** stigma; **G** gynoecium; **H** ovary; **J** habit. *S. crassifolia* subsp. *rotundifolia*. **K** fruit; **L** leaf. *S. crassifolia* subsp. *abairensis*. **M** leaf. **A** from *Furlan et al.* SPF 18966/CFCR 1589; **B – J** from *Melo Silva* 436a; **K & L** from *Arbo* 5438; **M** from *Fothergill* 131; **N** from *Harley et al.* 22667. DRAWN BY LINDA GURR.

(K); Campo da Pedra Grande, 2 March 1992, *Sano & Læssøe* H 50993 (K); Arredores de Catolés, 15 March 1992, *Stannard et al.* H 51983 (K); Campos Virasaia, 30 Dec. 1993, *Ganev* 2737 (K); Mun. Rio de Contas, Serra dos Brejões, 9 Aug. 1993, *Ganev* 2066 (K); Pico das Almas, 12 Nov. 1988, *Harley et al.* 26430 (K).

HABITAT. *Campo rupestre* at high altitudes (1250 – 1850 m), often in sand.

CONSERVATION STATUS. Least concern.

Queiroz 5031 may fit here. Collected Mun. Abaíra, Distrito de Catolés, estrada para Serrinha e Bicota, but this specimen has large (c. 10 × 3.5 cm), elliptic leaves.

c. subsp. **rotundifolia** *S. Atkins* **subsp. nov.** a subsp. *crassifolia* habitu frutescenti ramoso usque 1.5 m; foliis rotundatis (non ovatis) 1.5 – 4 × 1.2 – 3.2 cm differt. Typus: Bahia, Morro do Chapéu, 31 May, 1980, *Harley et al.* 22831 (holotypus SPF; isotypi CEPEC; K).

Stachytarpheta crassifolia forma *alba* Moldenke (1984b: 380). Type: Brazil, Bahia, Serra do Sincorá, 22 Jan. 1984, *Hatschbach* 47464 (holotype TEX; isotype MBM).

Shrub to 1.5 m. Leaves rotund, 1.5 – 4 × 1.2 – 2.8 cm. Fig. 8K, L.

DISTRIBUTION. Bahia: Mun. Morro do Chapéu (Map 6).

SPECIMENS EXAMINED. BAHIA: Mun. de Morro do Chapéu, Rio do Ferro Doido, 19.5 km SE of Morro do Chapéu, 1 March 1977, *Harley et al.* 19196 (K); c. 16 km ESE de Morro do Chapéu, Cachoeira do Ferro Doido, 28 Feb. 1992, *Arbo et al.* 5438 (K); Estrada do Morro do Chapéu-Feira de Santana, 22 Feb. 1993, *Amorim et al.* 1026 (K); 18 km de Morro do Chapéu, na estrada para Feira de Santana, 13 March 1996,

Roque et al. PCD 2331 (K); 28 June 1996, *Hind et al.* 3170 (K).

HABITAT. *Campo rupestre*, amongst rocks, on sand at 900 – 1100 m.

CONSERVATION STATUS. Least concern.

d. subsp. **minasensis** *S. Atkins* **subsp. nov.** a subsp. *crassifolia* in Minas Gerais (non Bahia) crescenti et foliis ovatis (non ellipticis usque anguste ovatis), inflorescentiis usque 28 cm longis (non 25 cm), tubo corollae usque 20 mm (non 16 mm) differt. Typus: Minas Gerais, Mun. Jequitinhonha, 20 Oct. 1988, *Harley et al.* 25235. (holotypus SPF; isotypus K).

Branched shrub to 1.5 m. Leaves ovate, 4 – 6.5 × 1.8 – 2.6 cm. Inflorescence up to 28 cm long by 6 mm wide; corolla tube to 20 mm. Fig. 8B – J.

DISTRIBUTION. Minas Gerais (Map 6).

SPECIMENS EXAMINED. MINAS GERAIS: Jequitinhonha, na descida para Pedra Azul, 11 June 1991, *Mello-Silva et al.* 436a (K).

HABITAT. *Campo rupestre* and *cerrado*, in sand at 1100 m.

CONSERVATION STATUS. Vulnerable. Only two collections have been made of this subspecies. The locality is away from the main massif of the Serra do Espinhaço, and any efforts at conservation of the Serra could miss this outlying *campo rupestre*. The paucity of specimens and the outlying nature of the locality could make this a vulnerable taxon.

Group 2 Gesnerioides
Six species in Brazil.
A group with longer, wider inflorescences, a woody rootstock, and a flattened fruit apex.

Key to species in the Gesnerioides group

1 Leaf base auriculate, clasping stem, with auricles overlapping those of opposite leaf · · · 22. **S. amplexicaulis**
  Leaf base not as above · 2
2 Calyx at anthesis with sinus between outer and inner teeth on abaxial side; bract glabrous with long
  uniseriate hairs along margin only · 3
  Calyx at anthesis with 4 ± regular teeth on abaxial side with no sinus between; bracts scabrid, without
  long uniseriate hairs along margin · 5
3 Stem distinctly winged; inflorescence with flowers somewhat laxly arranged, and rachis clearly visible · · · · · ·
  · 23. **S. alata**
  Stem not winged, or only indistinctly; inflorescence with flowers tightly packed, with rachis not visible · · · 4
4 Leaf apex acute, base narrowing to petiole, and only just wider than petiole, giving a petiolate
  appearance; corolla tube to 25 mm long, lobes to 3 mm across · · · · · · · · · · · · · · · · 24. **S. sprucei**
  Leaf apex obtuse, base narrowing to petiole, but much wider giving sessile appearance; corolla tube
  to 15 mm long, lobes to 6 mm across · 20. **S. gesnerioides**
5 Upper leaf surface almost glabrous; inflorescence often consisting of 2 or more spikes arising from
  same axis · 21. **S. reticulata**
  Upper leaf surface sparsely covered with uniseriate hairs; inflorescence of only 1 spike per axis · 25. **S. rupestris**

20. Stachytarpheta gesnerioides *Cham.* (1832: 245).
Schauer (1847: 566; 1851: 208); Dubs (1998). Type: "e
Brasilia", *Sellow* s.n. (holotype B† ; possible isotype K) .
S. azurea Moldenke (1940: 470). Type: Brazil, Mato
  Grosso, Porto Espiridião, Nov. 1908, *Hoehne* comm.
  Rondon 692 (holotype SP n.v.; photo NY).

Woody herb or subshrub to 2 m, with woody
underground rootstock, forming large clumps, often
unbranched, occasionally branching below
inflorescence. Stem markedly 4-sided, sometimes
hollow, sometimes slightly winged, sparsely to densely
covered with uniseriate hairs, more numerous on
opposite faces of stem and at leaf bases. Leaves sessile
to short-petiolate, patent, subcoriaceous, ovate to
broadly ovate, 5.5 – 12 × 2.8 – 6 cm, apex obtuse to
rounded, base attenuate-cuneate, decurrent into
petiole, upper surface somewhat bullate to rugose or
smooth, with nerves impressed, glabrous or sparsely
or densely covered with uniseriate hairs, lower
surface glabrous or sparsely or densely covered with
uniseriate hairs, with prominent reticulate nerves,
margin crenate. Inflorescence curving, 25 – 54 cm
long × 10 mm wide, occasionally previous season's
inflorescence present; bracts narrowly triangular,
woody, 12 mm long. Calyx straight, 10 mm long,
outer surface sparsely hairy with uniseriate hairs, with
5 short teeth divided by sinus on abaxial side. Corolla
dark blue with yellow throat, hypocrateriform, tube
straight, c. 15 mm long, lobes c. 6 mm across.
Stamens inserted at top of corolla tube. Ovary flask-
shaped, 3 mm long, style 25 mm long. Fruit greeny-
brown, surface slightly reticulate.

1. Upper and lower leaf surface glabrous · · · · · · · · ·
   · b. var. **glabra**
   Upper and lower leaf surface not glabrous · · · · 2
2. Upper and lower leaf surface with sparse scattered
       uniseriate hairs · · · · · · · · · a. var. **gesnerioides**
   Upper and lower leaf surface densely covered
       with uniseriate hairs · · · · · · · · · c. var. **hirsuta**

a. var. **gesnerioides**

Woody herb or subshrub to 2 m, with woody
underground rootstock, forming large clumps, often
unbranched. Leaves sessile to short-petiolate, patent,
broadly ovate, 6.5 – 12 × 3 – 6 cm, apex obtuse, base
attenuate, upper surface somewhat bullate/rugose,
with scattered uniseriate hairs, lower surface with
hairs along nerves only. Fig. 9A – E.

DISTRIBUTION. Bolivia, Paraguay & Brazil: Distrito
Federal, Goiás, Mato Grosso, São Paulo (Map 7).
SPECIMENS EXAMINED. DISTRITO FEDERAL: Fazenda Água
Limpa, near Vargem Bonita, 27 April 1976, *Ratter &*

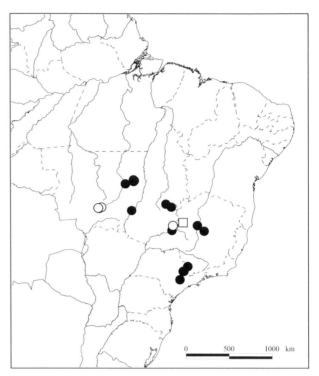

Map 7. Distributions of *Stachytarpheta gesnerioides* var.
gesnerioides (●), *S. gesnerioides* var. *glabra* (○) and
S. gesnerioides var. *hirsuta* (□).

Fonseca R2942 (K). **GOIÁS:** Inter Corumbá et
Corallinho, comm. 1839, *Pohl* s.n. (BR); 1896, *Glaziou*
21894 (K); c. 15 km S of Niquelândia, 21 Jan. 1972,
Irwin et al. 34652 (K); 25 km NE of Catalão, 21 Jan.
1970, *Irwin et al.* 25016 (K). **MATO GROSSO:** E of km
242, Xavantina to Cachimbo road, 4 Jan. 1968,
Philcox & Ferreira 3872 (K); c. 270 km N of Xavantina,
3 March 1968, *Gifford* G14 (K); 8 km NE of Base
Camp, 9 April 1968, *Ratter et al.* R910 (K); Xavantina
to Cachimbo road, 65 km from Xavantina, 24 May
1966, *Hunt* 5539 (K); Lagoa de Leo, c. 270 km N of
Xavantina, 8 May 1968, *Ratter et al.* R1372 (K); Rio
Sagaru, between Rio Amolar and Rio Nobres, June
1927, *Dorrien-Smith* 239 (K). **SÃO PAULO:** Mun. Moji-
Guaçu, Reserva Biol. da Fazenda Campininha, 26
Jan. 1981, *Mantovani* 1528 (K). **PARAGUAY:** Sierra de
Amambay, 1907/1908, *Rojas* 9802 (K). **BOLIVIA:** Santa
Cruz, 6 Nov. 1993, *Guillen & Centurion* 861 (K).
HABITAT. Dry, grassy slopes, *campos* and *cerrado*, up to
1000 m.

CONSERVATION STATUS. Least concern. Fairly common,
and found in areas of *cerrado* in many localities.

A tall plant with long spike and thick leaves.

b. var. **glabra** *S. Atkins* **var. nov.** a var. *gesnerioide* foliis
ubique glabris (non pilis uniseriatis sparsis obsitis),
pagina superiore laevi (non rugosa) distincta.

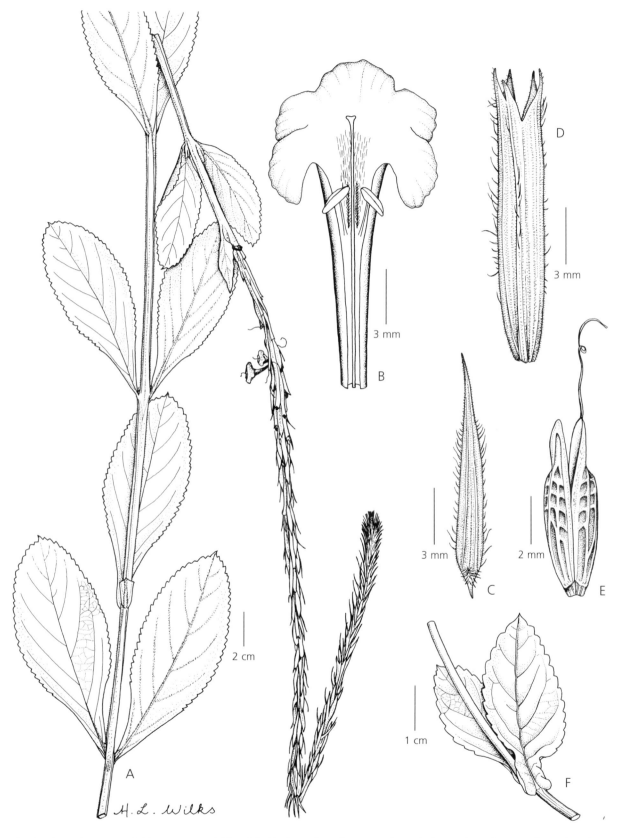

Fig. 9. *Stachytarpheta gesnerioides* var. *gesnerioides*. **A** habit; **B** corolla. **C** bract; **D** calyx; **E** fruit. *Stachytarpheta amplexicaulis*. **F** leaves showing base of leaf and leaf attachment. **A**, **B** from *Mantovani* 1528; **C**, **D** from *Onishi et al.* 877; **E** from *Anderson* 9672; **F** from *Hatschbach* 46311. DRAWN BY HAZEL WILKS.

Typus: Mato Grosso, camino a Águas Quentes, 30 Jan. 1989, *Krapovickas & Cristobal* 43149 (holotypus K; isotypus CTES).

Shrub or herb. Stems 4-sided, slightly winged, glabrous. Leaves sessile, erect, ovate, 7.5 – 12 × 3.5 – 5 cm, upper and lower surface glabrous.

DISTRIBUTION. Goiás and Mato Grosso (Map 7).
SPECIMENS EXAMINED. GOIÁS: Mun. Campo Alegre, BR 050, próximo Rio Imburuçu, 29 Nov. 1992, *Hatschbach & Barbosa* 58264 (K); MATO GROSSO: Cuiabá, Burity, Oct. 1927, *Collenette* 187 (K); Chapada dos Guimarães, 27 Jan. 1989, *Cavalcanti et al.* 92 (K).
HABITAT. In stony soil.
CONSERVATION STATUS. Data deficient. This variety occurs in Goiás and Mato Grosso. Paucity of specimens make this taxon difficult to assess.

c. var. **hirsuta** *S. Atkins* **var. nov.** a var. *gesnerioide* caulibus foliisque pilis uniseriatis densius obsitis differt. Typus: Mato Grosso: km 99 da rodovia Cuiabá – Porto Velho, 24 Feb. 1982, *Santos & Rosario* 503 (holotypus INPA; isotypus K).

Herb or subshrub to 1 m. Stems 4-sided covered with uniseriate hairs, pointing in all directions. Leaves sessile, erect, ovate, 5.5 – 8.5 × 2.8 – 5 cm, base cuneate, apex obtuse, sometimes rounded, upper and lower surface fairly densely covered with uniseriate hairs, margin serrate.

Only known from the type (Map 7).

21. Stachytarpheta reticulata *Mart. ex Schauer* (1847: 566; 1851: 209). Type: Minarum Generalium, 'in campis das Lavras de Pindaíba, Serro Frio, itemque in summo Brasiliae monte Itambé: Mart. (Herb. Bras. a Claussen coll. 1840. no. 386)'. (isosyntype BR).
S. irwinii Moldenke (1971: 254). Type: Minas Gerais, Serra do Espinhaço, 3 km N of São João Chapada, 24 March 1970, *Irwin et al.* 28208 (holotype TEX).
S. lacunosa Schauer var. *cordifolia* Moldenke (1973a: 221). Type: Minas Gerais, Serra do Espinhaço, 30 km NE Francisco Sá, 1100 m, 10 Feb. 1969, *Irwin et al.* 23029 (holotype NY).
S. lacunosa var. *ovatifolia* Moldenke (1973a: 222). Type: Minas Gerais, Serra do Espinhaço, 48 km W of Montes Claros, 950 m, 25 Feb.1969, *Irwin et al.* 23871 (holotype TEX; isotype NY).
S. lacunosa var. *attenuata* Moldenke (1974a: 304). Type: Minas Gerais, Serra do Espinhaço, 12 km SW of Diamantina, 1370 m, 18 Jan. 1969, *Irwin et al.* 22157 (holotype TEX; isotype NY).

Shrub to 2 m. Stems rounded, pubescent. Leaves sessile, erect or patent, ovate to subrotund, 2.5 – 5 (7) × 1.3 – 2 (4) cm, apex acute to obtuse, base cuneate or truncate or cordate, margin crenate or serrate, upper surface almost glabrous or with fine scattering of short hairs, lower surface with veins forming a reticulate network with short hairs along veins. Inflorescence 3.5 – 33 cm long × 10 mm wide (inflorescence often with 1 or more branches arising from same axis, or new inflorescence branch arising from just below previous year's inflorescence), bracts narrowly triangular, 6 mm long. Calyx 10 mm long with 4 teeth on abaxial side. Corolla hypocrateriform, dark blue, tube straight, c. 15 mm long, lobes c. 6 mm across. Stamens attached above middle of tube, anthers lying just below throat. Ovary narrow, c. 2 mm long. Fruit light brown, with prominent attachment scar, 7 mm long. Fig 10G – M.

DISTRIBUTION. Minas Gerais (Map 8).
SPECIMENS EXAMINED. MINAS GERAIS: Without locality, 1837, Herb. Mus. Vind. 174 (K); without locality, 1841, *Gardner* 5111 & 5112 (K); Serra do Lenheiro et Biribiry, prés Diamantina, 1892, *Glaziou* 19719 (K); Serra do Espinhaço, 18 Jan. 1969, *Irwin et al.* 22157 (NY); 10 Feb. 1969, *Irwin et al.* 23029 (NY); 25 Feb. 1969, *Irwin et al.* 23871 (K); Mun. Joaquim Felício, Serra do Cabral, 6 March 1970, *Irwin et al.* 27051 (K); 16 April 1981, *Rossi et al.* CFCR 1042/SPF 22968 (K);

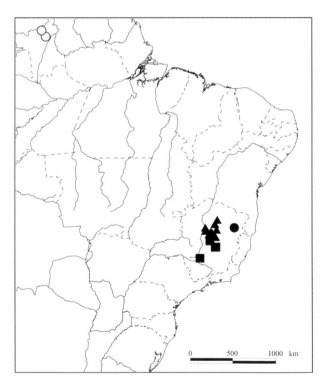

Map 8. Distributions of *Stachytarpheta amplexicaulis* (●), *S. reticulata* (▲), *S. rupestris* (■) and *S. sprucei* (○).

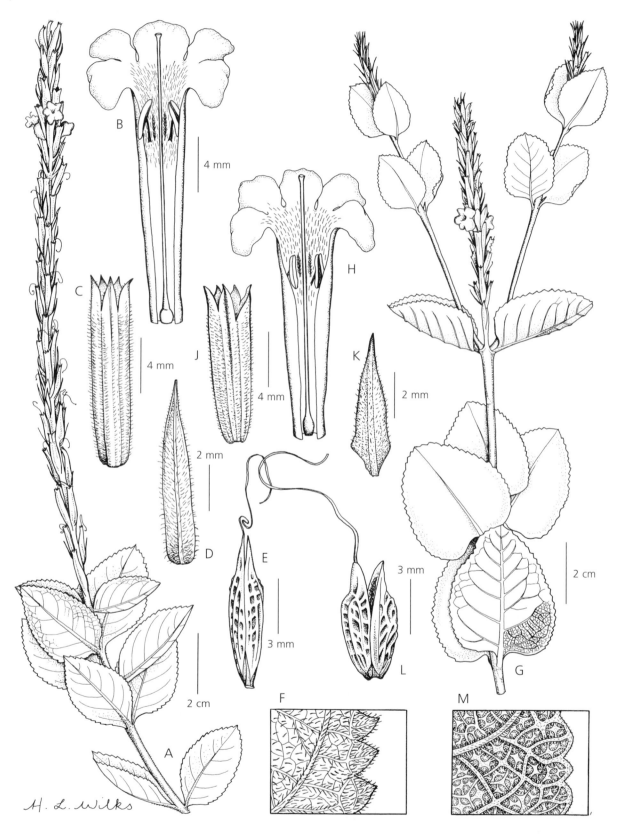

Fig. 10. *Stachytarpheta rupestris.* **A** habit; **B** corolla; **C** calyx; **D** bract; **E** fruit; **F** detail of underside of leaf. *Stachytarpheta reticulata.* **G** habit; **H** corolla; **J** calyx; **K** bract; **L** fruit; **M** detail of underside of leaf . **A – E** from *Atkins et al.* CFCR 13932; **F** from *Krapovickas & Cristobal* 33418; **G, M** from *Hatschbach* 61814; **H – K** from *Irwin et al.* 23871; **L** from *Irwin et al.* 27051. DRAWN BY HAZEL WILKS.

Mun. Buenopolis, Serra do Cabral, 13 Oct. 1988, *Harley et al.* 24922 (SPF); 19 March 1994, *Roque et al.* CFCR 15193 (K); Mun. Augusto de Lima, Serra do Cabral, 20 March 1994, *Roque et al.* CFCR 15308 (K); Mun. Grão Mogol, Serra do Calixto, 14 Oct. 1988, *Harley et al.* 25009 (K); Mun. Serro, 10 Aug. 1972, *Hatschbach* 30152 (NY); Mun. Felisberto Caldeira, 16 Feb. 1972, *Hatschbach & Ahumada*, 31674 (NY).

HABITAT. *Campo rupestre* and *cerrado*, 850 – 1200 m.

CONSERVATION STATUS. Least concern.

Moldenke (1973a, 1974a) seems to have confused *Stachytarpheta reticulata* with Schauer's *S. lacunosa*. Certainly Schauer's description 'leaves thick, hard.....scrobiculate-reticulate beneath with a strongly reticulate sculpturing' could apply to *S. reticulata*, but *S. lacunosa* does not have the elongated fruit apex, has smaller leaves, shorter inflorescence and a much larger corolla limb, and is confined to the Chapada around the town of Rio de Contas in Bahia.

22. Stachytarpheta amplexicaulis *Moldenke* (1947a: 320). Type: Minas Gerais, Congonhas do Campo, *Glaziou* 13063 (isotypes G, K).

Subshrub. Stems rounded, glabrous to very sparsely covered with short, white hairs. Leaves sessile, ovate, 3.5 – 8 cm, apex acute, base auriculate, clasping stem, with auricles overlapping those of opposite leaf, margin crenate, upper surface almost glabrous, lower surface puberulent, with simple hairs. Inflorescence 8 – 15 cm long × 10 – 15 mm wide; bracts narrowly triangular, c. 10 mm long. Calyx straight, ribbed c. 13 mm long, with 2 sinuses, and 4 (?5) twisted teeth. Corolla hypocrateriform, blue, tube c. 20 mm long, lobes c. 4 mm across. Stamens attached above middle of tube, anthers lying at throat. Ovary 2 mm long, style 20 mm long. Fruit pale, c. 8 mm long. Fig. 9F.

DISTRIBUTION. Minas Gerais (Map 8).

SPECIMENS EXAMINED. MINAS GERAIS: Mun. Catuji, Pontelete, 13 May 1983, *Hatschbach* 46311 (MBM, NY).

HABITAT. Disturbed or secondary vegetation (Capoeira).

CONSERVATION STATUS. Data deficient. Only one modern collection. Catuji is right next to the main road between Bahia and Rio de Janeiro. There is insufficient data, but the taxon could be under threat.

This is easily distinguishable by its overlapping leaf bases.

23. Stachytarpheta alata (*Moldenke*) *S. Atkins* **stat. nov.** Type: Minas Gerais, Mun. Medina, 13 May 1983, *Hatschbach* 46323 (holotype MBM; isotype G).

S. gesnerioides Cham. var. *alata* Moldenke (1983b: 399). Type as for species.

Subshrub to 50 cm. Stem 4-sided, winged. Leaves petiolate, patent, ovate, 2.5 – 7 × 1.4 – 3 cm, apex obtuse, base attenuate, decurrent into petiole, margin crenate, upper and lower surface covered with uniseriate hairs, more dense on lower surface. Inflorescence 10 cm × 15 mm wide; bracts narrowly triangular, c. 9 mm long, ± more or less glabrous with scattered uniseriate hairs along margin. Calyx with 5 short teeth and sinus on abaxial side. Corolla hypocrateriform, blue, tube c. 18 mm long, lobes c. 4 mm across. Stamens attached above middle of tube. No fruit. Fig. 11F – H.

DISTRIBUTION. Minas Gerais.

HABITAT. Among rocks.

Known only from the type.

CONSERVATION STATUS. Data deficient.

Similar to *Stachytarpheta gesnerioides*, this species differs in its smaller and less coriaceous to bullate leaves, and its shorter and more lax inflorescence. I have found no fruit, and place the species here tentatively.

24. Stachytarpheta sprucei *Moldenke* (1940: 474). Jansen-Jacobs (1988). Type: Venezuela, Rio Orinoco, near Maypures, June 1854, *Spruce* 3631 (holotype NY; isotype K).

Shrub to 4 m. Stem 4-sided, sparsely hairy. Leaves petiolate, patent, ovate, 6.5 – 16 × 2.7 – 8 cm, apex acute, base attenuate, decurrent into petiole, margin crenate, upper and lower surface scattered with uniseriate hairs, longer and more numerous on lower surface. Inflorescence up to 30 cm long × 10 mm wide; bracts subulate, c. 11 mm long, sparsely hairy, but with long uniseriate hairs along margin. Calyx c. 16 mm long, sparsely hairy along nerves, 4-dentate, with 2 sinuses. Corolla blue, hypocrateriform, tube with kink, c. 25 mm long, lobes c. 3 mm across, pubescent at throat. Stamens inserted towards top of tube. Ovary ovoid, c. 3 mm long, style c. 23 mm long. Fruit pale brown, beaked, c. 9 mm long without prominent attachment scar. Fig. 11A – E.

DISTRIBUTION. Venezuela, Guyana and Brazil (Roraima) (Map 8).

SPECIMENS EXAMINED. VENEZUELA. BOLIVAR: Serra do Sol, 31 Dec. 1954, *Maguire & Maguire* 40460; Cerro Negro Parado, 27 Dec. 1955, *Wurdack & Monachino* 40980. **BRAZIL. RORAIMA:** Serra de Pacaraima, Nov. 1909, *Ule* 8295 (K); Rio Branco, Surumu, Feb. 1909, *Ule* 7975 (K).

HABITAT. In savanna at 100 – 350 m.

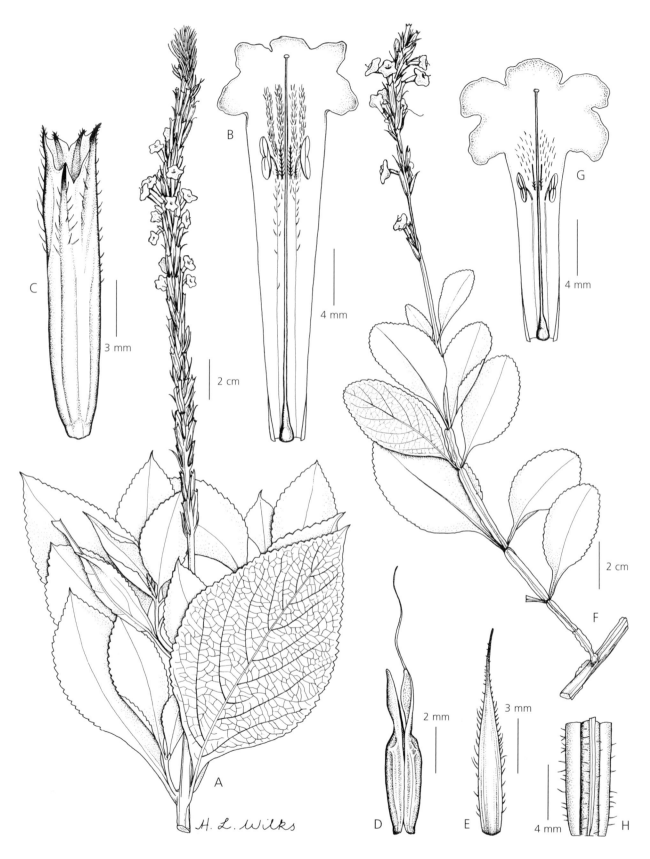

Fig. 11. *Stachytarpheta sprucei.* **A** habit; **B** corolla; **C** calyx; **D** fruit; **E** bract. *Stachytarpheta alata.* **F** habit; **G** corolla; **H** detail showing winged stem. **A** from *Ule* 7875; **B – E** from *Ule* 8295; **F – H** from *Hatschbach* 46323. DRAWN BY HAZEL WILKS.

CONSERVATION STATUS. Data deficient. The distribution spans the borders of Venezuela, Brazil and Guyana and is part of the "Transverse Dry Belt" (Pires-O'Brien 1997). This includes an area of lowland savanna forest. A northern east-west highway was planned by the Brazilian government in 1973, but has not been completed due to the lack of financial resources.

Without doubt, this species belongs here, because of the morphology of the fruit and the form of the inflorescence. Its longer corolla tube, and smaller lobes make it different from the other species in the group.

25. Stachytarpheta rupestris *S. Atkins* **sp. nov.** *S. gesnerioides* caulibus ± teretibus (non valde 4-lateralibus), foliis usque 4 cm longis (non usque 12 cm) supra laevibus (non bullatis), calyce latere adaxiali sinu carenti distinguenda differt. Typus: Minas Gerais, estrada Juscelino Gouveia, 9 km de Presidente Juscelino, 9 Feb. 1994, *Atkins et al.* CFCR 13932 (holotypus SPF; isotypus K).

Shrub to 75 cm. Stem 4-sided, fairly densely covered in short, white uniseriate hairs. Leaves, patent, sessile, subcoriaceous, ovate, 3 – 4 × 1.8 – 2.2 cm, apex obtuse, base cuneate, margin serrate, upper surface sparsely covered with uniseriate hairs, and scattered with small brown sessile nectaries, lower surface with reticulate network of veins, and densely covered with uniseriate hairs. Inflorescence 16 – 35 cm long × c. 7 mm wide; bracts narrowly triangular, 9 mm long, with fine covering of short, white hairs. Calyx c. 14 mm long, with 4 abaxial teeth and 1 adaxial. Corolla infundibular, brilliant blue with white throat, tube c. 18 mm long, lobes c. 5 mm

across. Stamens attached above middle of tube, anthers lying towards throat. Ovary 1 mm long, style 18 mm long. Fruit yellow, with reticulate surface. Fig. 10A – F, Plate 6C.

DISTRIBUTION. Minas Gerais (Map 8).
SPECIMENS EXAMINED. MINAS GERAIS: Near Formigas, July 1840, *Gardner* 5109 (K); Serra do Lenheiro et Biribiry (Biribiri), without date, *Glaziou* 17714a (K); Lagoa Santa, 2 Feb. 1978, *Krapovickas & Cristobal* 33418 (K); Mun. Santana do Riacho, 13 Feb. 1984, *Longhi-Wagner et al.* CFCR 5927/SPF 34987 (K).
HABITAT. In *cerrado* and *campo rupestre* at 820 m; often on rock.
CONSERVATION STATUS. Vulnerable. Modern collections form a triangle north of Belo Horizonte, covering an area of about 25,000 km^2. None of this area is currently protected; there are major conurbations close by, and some areas are grazed and planted.

A plant with restricted distribution. It has a much longer spike than *Stachytarpheta reticulata,* and smaller leaves than *S. gesnerioides.*

Group 3 Microphylla
Two species in Brazil.
These two species are difficult to place. They differ from the Cayennensis group by the slightly drawn-out and flattened fruit apex. This character should put them with the Gesnerioides group, but they differ by their deeply excavated rachis, with the calyx submerged in the excavations, as in the Cayennensis group. *Stachytarpheta microphylla* has a woody rootstock, but *S. sessilis* does not. The calyx is different from both groups in that it is distinctly bilobed. Both species have a bright red corolla. I put them here in a separate group of their own.

Key to species in the Microphylla group

1. Stem with dense, spreading hairs continuing onto base of leaves; normally more than one pair of leaves at each axil; rachis ± glabrous, excavations in fruiting rachis narrower than rachis · · · · **26. S. microphylla**
   Stem with sparse, spreading hairs; normally only one pair of leaves at each axil; rachis covered with hairs; excavations in fruiting rachis as broad as rachis · **27. S. sessilis**

26. Stachytarpheta microphylla *Walp.* (1844: 6). Type: Bahia, Jacobina, 1840, *Blanchet* 3120 (neotype K, **selected here**).
Stachytarpheta sanguinea Schauer & Mart. in Schauer (1847: 564). Type: Bahia, Jacobina, 1840, *Blanchet* 3120 (isosyntype K).
Stachytarpheta sanguinea var. *hatschbachii* Moldenke (1983c: 67). Type: Bahia, Mun. Oliveira dos

Brejinhos, 12 Oct. 1981, *Hatschbach* 44170. (holotype TEX; isotype MBM).

Subshrub to 1 m, branched. Stems woody, rounded, covered with long white spreading uniseriate hairs, more numerous at nodes, and spreading up on to base of leaves. Leaves sessile, somewhat clasping, often several smaller leaves in the axils, oblong, 1.2 –

4 × 0.6 – 2.2 cm, apex acute, base acuminate, margin regularly to irregularly crenate-serrate, upper and lower surface villous to densely villous with uniseriate hairs. Inflorescence 10 – 26 cm long by 3 – 5 mm wide, rachis ± glabrous; bracts woody, narrowly triangular, c. 10 mm long. Calyx 12 mm long, 2-lobed, with deep sinus, dark brown sessile glands on calyx teeth. Corolla hypocrateriform, red, tube 14 mm long, lobes c. 6 mm across. Stamens lying just below throat, attached towards top of tube. Ovary flask-shaped, c. 1.5 mm long. Fruit very deeply embedded in rachis, fruit blackish with stylopodium and flattened apex.

DISTRIBUTION. Bahia and Piauí (Map 9).
SPECIMENS EXAMINED. BAHIA: Mun. Entre Rios, W. of Subaúma, 28 May 1981, *Mori & Boom* 14157 (NY, K); Mun. Rio de Contas, caminho para Lagoa Nova, 5 Feb. 1997, *Passos et al.* PCD 5110 (K); Serra do Tombador, 17 Jan. 1997, *Arbo et al.* 7388 (NY). **PIAUÍ:** without locality, 1839, *Gardner* 2286 (K).
HABITAT. Disturbed vegetation, roadside, close to *caatinga*.
CONSERVATION STATUS. Least concern. Although it has a fairly narrow distribution, this species is reasonably common, and can quickly colonise waste ground.

This differs from all preceding species in its bright red flowers. It differs from *Stachytarpheta sessilis* by the density of the indumentum on the stem, by the glabrous rachis, and by the excavations in the rachis being narrower than the rachis (not the same width as in *S. sessilis*).

Although Walpers' name *Stachytarpheta microphylla* predated Schauer's, Schauer put it into the synonymy of his own *S. sanguinea*. This name has been used since then for this species. Schauer stated that Walpers' specimen was not adequate. Walpers did not cite type specimens for his new species. His herbarium was sold after his death, and its whereabouts is not known. Schauer said that the specimen he saw was in Herb. Lucca (Herb. Lucaeani), but this herbarium has no such specimen. It is clear from Walpers' original description (protologue) that the taxon he described is conspecific with *S. sanguinea*. I have chosen to neotypify Walpers' name, using a syntype cited by Schauer, i.e. *Blanchet* 3120. There are two specimens at Kew, in good condition, numbered and annotated 'Jacobina'.

27. Stachytarpheta sessilis *Moldenke* (1947b: 371). Type: Ceará, Salvação, 6 March 1910, *Löfgren* 160 (holotype S n.v.; Photo NY).

Herb to 50 cm, unbranched. Stems slender, 4-sided, canaliculate, lightly covered with spreading,

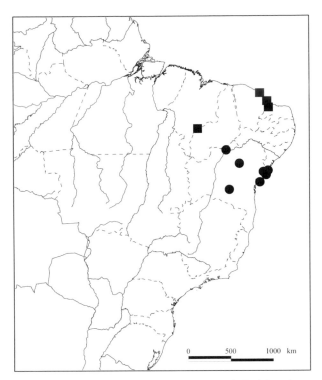

Map 9. Distributions of *Stachytarpheta microphylla* (●) and *S. sessilis* (■).

uniseriate hairs. Leaves sessile, patent, oblong, 2.2 – 3.5 × 0.4 – 1.8 cm, apex acute, base truncate, margin irregularly serrate, upper and lower surface sparsely covered with uniseriate hairs. Inflorescence 9 – 19 cm long × 5 mm wide, rachis hairy; bracts narrowly triangular, c. 8 mm long. Calyx 9 mm long with 2 teeth and sinus. Corolla infundibular, deep red, tube slightly curved, c. 10 mm long, lobes c. 5 mm across. Stamens inserted at middle of tube. Ovary c. 1 mm long, style c. 6 mm long, included. Fruit pale, apex drawn out and flattened.

DISTRIBUTION. Ceará, Maranhão, Rio Grande do Norte (Map 9).
SPECIMENS EXAMINED. CEARÁ: Aracati (Aracaty), July 1838, *Gardner* 1815 (K); Fortaleza, Campus do Pici, *Cavalcanti* s.n. (K). **MARANHÃO:** Mun. Loreto, Ilha de Balsas, 24 March 1962, *Eiten & Eiten* 3776 (K). **RIO GRANDE DO NORTE:** Mun. Baraúna, 31 May 1984, *Collares & Dutra* 138 (K).
HABITAT. *Caatinga* and open "*tabuleiro*".
CONSERVATION STATUS. Near threatened. Very few collections. The *caatingas* of these areas are now isolated from each other and some are now within urban areas.

Similar to *Stachytarpheta microphylla*, but more delicate, and with leaves not irregularly lobed.

Group 4 Quadrangula
Seven species in Brazil.

Key to species in the Quadrangula group

1. Flowers white · 33. **S. galactea**
   Flowers red or blue · 2
2. Flowers red · 3
   Flowers blue · 4
3. Leaves almost completely glabrous; upper surface scabrid; inflorescence to 5 mm wide (excluding open
      flower); bracts ± subulate, with parallel sides at base, and then tapering upwards from the midpoint,
      glabrous, striate, c. 15 mm long · 30. **S. scaberrima**
   Leaves covered in fine layer of short, uniseriate hairs; upper surface not scabrid; inflorescence to 12 mm
      wide (excluding open flower); bracts narrowly triangular, apex attenuate, covered with uniseriate
      hairs, 10 mm long · 32. **S. coccinea**
4. Inflorescence 2 – 5 mm wide (excluding open flower); calyx almost glabrous (scattered long hairs
      along nerves at most) · 5
   Inflorescence 8 – 10 mm wide (excluding open flower); calyx with long uniseriate hairs, especially
      along nerves · 6
5. Stem markedly 4-sided with dark vertical line at each edge; bracts c. 3 – 7 mm long at anthesis with
      apex short-acuminate; corolla tube to 18 mm long · · · · · · · · · · · · · · · · · 28. **S. quadrangula**
   Stem rounded to 4-sided; bracts c. 12 –15 mm long at anthesis with apex long-acuminate; corolla tube
      c. 30 mm long · 29. **S. bicolor**
6. Inflorescence normally of 3 similar spikes (1 terminal plus 2 in axils of uppermost leaf pair; see Fig. 14);
      bract keeled; calyx 12 mm long; corolla tube to 25 mm long · · · · · · · · · · · · · · · 31. **S. trispicata**
   Inflorescence with only single terminal spike without laterals in axils of uppermost leaf pair; bract
      without keel; calyx 22 mm long; corolla tube 40 mm long · · · · · · · · · · · · · · · · 34. **S. speciosa**

---

28. Stachytarpheta quadrangula *Nees & Mart.* (1823: 69). Schauer (1847: 567; 1851: 206). Type: Brazil, Bahia, Tamburil and Valos, ?date, *Prince Maximilian of Wied*, (holotype M n.v.; isotype BR).

Shrub or treelet to 2 m, much branched. Stems woody, 4-sided, with black edge at each corner, sparsely hairy to glabrous. Leaves petiolate, thickly chartaceous, narrowly ovate to ovate, 2 – 4 × 0.8 – 2 cm, apex acute, base attenuate, decurrent into petiole, margin crenate, upper and lower surface with minute sessile glands, sparsely hairy. Inflorescence up to 18 cm long × 2 – 3 mm wide, bracts narrow, 3 – 7 mm long, with short-acuminate apex. Calyx 14 mm long, bifid, lobes equal, 2-toothed, almost glabrous, with only minute hairs interspersed with occasional longer hairs. Corolla deep blue, narrowly infundibular, tube c. 18 mm, inner surface with hairs at throat, outer surface glabrous, lobes small, c. 3 mm across. Stamens inserted at top of tube, included. Ovary flask-shaped, c. 2 mm, style c. 26 mm long. Fruit light brown, surface smooth, with prominent attachment scar, apex elongated. Fig. 12, Plate 8A.

DISTRIBUTION. Bahia, Chapada Diamantina (Map 10).
SPECIMENS EXAMINED. BAHIA: Serra Açuruá, 1838, *Blanchet* 2820 (E; G; K); Andaraí, 24 Jan. 1980, *Harley et al.* 20558 (K); Catolés, 2 May 1992, *Ganev* 209 (K); 27 July 1992, *Ganev* 748 (K); 28 Nov. 1992, *Ganev* 1576 (K); Serra do Sincorá, 13 Feb. 1977, *Harley et al.* 18597 (K); 26 March 1980, *Harley et al.* 20964 (K); Contendas do Sincorá, 21 Nov. 1985, *Hatschbach & Zelma* 50074 (K); Mucugê, 22 Nov. 1996, *Hind et al.* PCD 4567 (K); Rio de Contas, 24 Jan. 1994, *Ganev* 2869 (K); 4 March 1994, *Sano et al.* CFCR 14878 (K).
HABITAT. Transition from *cerrado-caatinga*, *caatinga*, *campo rupestre*, semi-deciduous woodland at altitudes 400 – 1500 m.
CONSERVATION STATUS. Least concern. Although confined to Bahia, a relatively common species where found.

29. Stachytarpheta bicolor *Hook. f.* (1865: t. 5538). Type: "Bahia, Williams 2/1865 Bot. Mag. 5538" (K).
S. bicolor f. *pilosula* Moldenke (1981a: 255). Type: Bahia, Mun. de Conceição da Feira, 17 Feb. 1981, *Carvalho* 539 (holotype TEX).

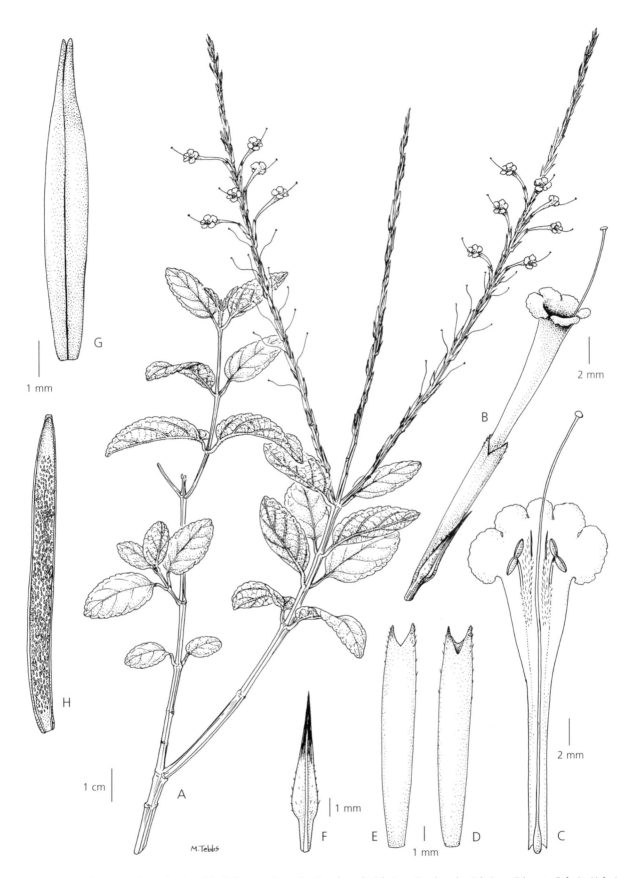

Fig. 12. *Stachytarpheta quadrangula*. **A** habit; **B** flower; **C** corolla; **D** calyx adaxial view; **E** calyx abaxial view; **F** bract; **G** fruit; **H** fruit dissected along commissure. **A**, **G**, **H** from *Hatschbach* 50074; **B – F** from *Harley* 20964. DRAWN BY MARGARET TEBBS.

Map 10. Distributions of *Stachytarpheta quadrangula* (●) and
S. scaberrima (■).

HABITAT. *Caatinga* and open scrub at 500 – 600 m.
CONSERVATION STATUS. Vulnerable. Not a common
species and some localities now grazed by cattle.

Very close to *Stachytarpheta quadrangula*, but confined
to an area around Itaberaba and Milagres. *S. bicolor*
has a longer corolla tube and bract, and lacks the
distinctive dark lines on the stem ridges.

30. Stachytarpheta scaberrima *Cham.* (1832: 244);
Schauer (1847: 567; 1851: 207). Type: 'e Brasilia'
Sellow s.n. (holotype B†; isotypes BR; G).

Shrub or treelet to 2 m. Stems markedly 4-sided, pale,
somewhat striate, glabrous except for a fringe of short
hairs at nodes. Leaves sessile to petiolate, patent,
herbaceous, ovate, 3 – 8 × 1.5 – 3.2 cm, apex acute,
base attenuate, decurrent into petiole, margin crenate,
upper and lower surface almost completely glabrous,
upper surface slightly scabrid. Inflorescence slightly
curving, 8 – 14 cm long × c. 5 mm wide; bracts ±
subulate, with parallel sides at base, and then tapering
from half the length, glabrous, striate, c. 15 mm long.
Calyx c. 18 mm long, bifid, with deep sinus, almost
glabrous. Corolla crimson red, hypocrateriform, tube c.
29 mm long. Stamens inserted at top of corolla tube,
anthers lying just below throat. Ovary pyriform, c. 2
mm long, style c. 32 mm long, exserted. Fruit dark
brown, narrow, 13 mm long, beaked. Fig. 13.

DISTRIBUTION. Bahia, Espírito Santo and Minas Gerais
(Map 10). The three collection areas form a triangle,
and it is possible that the species exists in other areas
inside the triangle, an area which has not been
collected.
SPECIMENS EXAMINED. BAHIA: Jequié, 22 Jan. 1965,
Belém & Mendes, 198 (K); Jequié, Bairro Suiça, 13
Feb. 2003, *França et al.* 4342; 13 Feb. 2003, *França et
al.* 4344. ESPÍRITO SANTO: Victoria, *Sellow* sn (K). MINAS
GERAIS: Mun. Almenara, Morro do Cruzeiro, 15 Feb.
1988, *Thomas et al.* 5990 (NY); Two *Sellow* specimens
without number and without locality (K).
HABITAT. Transition from *caatinga*. Specimens from
Jequié growing in degraded scrub.
CONSERVATION STATUS. Vulnerable. Not a widespread
or common species, and the area in which we
collected, just outside the town of Jequié, was being
heavily collected for firewood.

This species differs from *Stachytarpheta coccinea* by its
narrower inflorescence, and by its glabrous, scabrid
leaves. The specimens from around Jequié have
narrower bracts with ciliate margins.
  Glaziou 16290, Minas Gerais, Itacolumi, appears to
be *Stachytarpheta scaberrima* but the label data says
"fleures bleues".

Subshrub or shrub to 2 m, much branched. Stems
rounded to 4-sided, pale, almost glabrous,
lenticellate. Leaves petiolate, patent, herbaceous,
ovate, 3 – 7 × 1.4 – 3.8 cm, apex acute, base attenuate
to cuneate, decurrent into petiole, upper surface
almost completely glabrous, with some scattered
hairs along immersed veins, slightly scabrid, lower
surface slightly more hairy, with nerves slightly raised
and pale, with occasional glands on undersurface.
Inflorescence 8 – 22 cm long × 4 – 5 mm wide; bracts
subulate to narrowly triangular c. 12 – 15 mm long at
anthesis with apex long-acuminate. Calyx 20 mm
long, glabrous, deeply bifid, lobes equal without
teeth. Corolla blue, narrowly infundibular, tube c. 30
mm long, outer surface with glandular hairs, with
sessile glands at throat, but no hairs, lobes small, c. 1
mm across. Stamens lying at throat. Ovary flask-
shaped, c. 1 mm long, style 35 mm long, exserted.
Fruit light brown, c. 8 mm long, beaked and without
prominent attachment scar.

DISTRIBUTION. Bahia, area between Feira de Santana,
Milagres and Itaberaba (Map 11).
SPECIMENS EXAMINED. BAHIA: Anguera, 17 Sept. 1994,
França & Melo 1058 (HUEFS, K); Itaberaba, 10
March 1982, *de Oliveira* 413 (EPABA, K); on Iaçu
road, 24 Jan. 1980, *Harley et al.* 20526 (K); Morro de
Nossa Senhora dos Milagres, 6 March 1977, *Harley et
al.* 19454 (K).

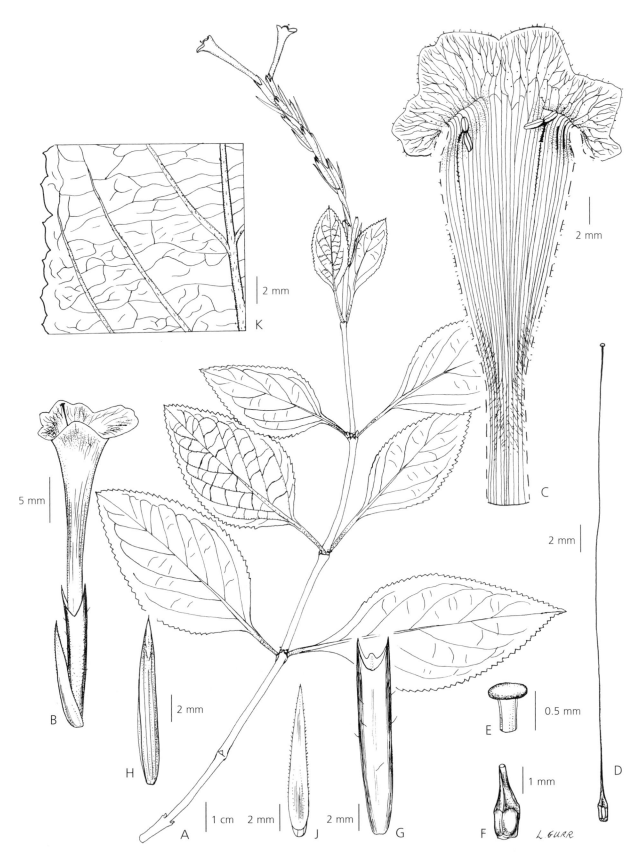

Fig. 13. *Stachytarpheta scaberrima*. **A** habit; **B** calyx and corolla; **C** corolla; **D** ovary, style & stigma; **E** apex of style; **F** ovary; **G** calyx; **H** fruit; **J** bract; **K** detail of leaf undersurface. **A** & **K** from *Belém & Mendes* 198; **B** – **J** from *França et al.* 4344. DRAWN BY LINDA GURR.

Map 11. Distributions of *Stachytarpheta bicolor* (■) and *S. coccinea* (●).

31. Stachytarpheta trispicata *Nees & Mart.* (1823: 70). Schauer (1847: 568; 1851: 211). Type: Brazil, Bahia, Tamburil et Valos, *Prince Maximilian of Wied* s.n. (holotype M n.v.; isotype BR).

S. trispicata var. *ovatifolia* Moldenke (1977: 408). Type: Brazil, Bahia, 41 km de Vitória da Conquista, 26 Jan. 1965, *Pereira* 9754 (isotype TEX).

Shrub or treelet to 3 m, much branched. Stems rounded, lenticellate, very pale, glabrous with scattered black glands. Leaves chartaceous, broadly ovate, apex acute, base attenuate, decurrent into petiole, petiole darker than stem, margin crenate-serrate, 5 – 9 × 2 – 4 cm, upper surface almost glabrous, lower surface with prominent pale veins, almost glabrous except for occasional hairs along nerves. Inflorescences normally 3 together, curving, 7 – 21 cm long × 8 – 10 mm wide; bracts narrow, long-attenuate at apex, keeled, margin scarious, with scattered uniseriate hairs along margin, c. 12 mm long. Calyx straight, c. 12 mm long, outer surface with scattered uniseriate hairs along nerves, 2-lobed, lobes 2-toothed, but with teeth often wrapped together. Corolla blue, infundibular, c. 25 mm long, lobes small, c. 2 mm across. Stamens inserted about half way up corolla tube, anthers lying towards top of corolla tube. Ovary flask-shaped, glabrous, c. 3 mm long, style glabrous, c. 30 mm long. Fruit with beak, c. 10 mm long. Fig. 14, Plate 5D.

DISTRIBUTION. Bahia.
SPECIMENS EXAMINED. BAHIA: Chapadão Occidental da Bahia, 24 April 1980, *Harley et al.* 21692 (K); Ituaçu, Morro da Mangabeira, 22 Dec. 1983, *Gouveia* 43/83 (K); Caetité, 10 March 1994, *Souza et al.* 5443 (K); Bom Jesus de Lapa, 20 April 1980, *Harley et al.* 21579 (K).
HABITAT. Secondary vegetation; in *caatinga* and dry deciduous forest.
CONSERVATION STATUS. Least concern. A reasonably common plant where it occurs.

Overall very similar to *Stachytarpheta coccinea*, but with blue flowers and with the inflorescences normally in groups of three.

32. Stachytarpheta coccinea *Schauer* (1847: 567; 1851: 207); Zappi *et al.* (2003). Type: Bahia, *Prince Maximilian of Wied* s.n. (syntype M n.v.; isosyntype BR); Jacobina, *Blanchet* 3885 (syntype G; isosyntype BR; photo NY).

S. gardneriana Hayek (1907: 274). Type: Ceará, Between Olho d'Agua & Poço do Cavalo, February 1839, *Gardner* 2434 (holotype W; isotype K).

S. loefgreni Moldenke (1947b: 369). Type: Ceará, Ingazeiro, 26 April 1910, *Löfgren* 692 (holotype S; fragment NY).

S. sanguinea var. *grisea* Moldenke (1981b: 330). Type: Bahia, Mun. de Maracás, 17 Nov. 1978, *Mori et al.* 11102 (isotype TEX).

Shrub or treelet to over 2 m, much branched. Stems woody, distinctly 4-sided, opposite faces slightly rounded, covered in short, white hairs. Leaves petiolate, sub-coriaceous, ovate, 2 – 6.5 × 1 – 4 cm, apex acute, base attenuate, decurrent into petiole, margin crenate to serrate, upper surface slightly rugose and covered in short, white, uniseriate hairs, lower surface with reticulate network of veins, lower surface more hairy. Inflorescence long, straight, 4 – 18 cm long × 7 – 15 mm wide; bracts narrowly triangular, apex attenuate, 10 mm long, covered with uniseriate hairs. Calyx straight, c. 14 mm long, 2-lobed, each lobe with 2 teeth, hairy on outer surface, inner surface glabrous. Corolla bright red, infundibular, tube slightly curved, 30 mm long, lobes small, c. 4 mm across, all ± equal, glabrous on outer surface, inner throat with hairs and glands. Stamens inserted just below throat. Ovary ovoid, 2 mm long, style c. 35 mm long. Fruit light brown, smooth, beaked. Plate 8C.

DISTRIBUTION. Bahia, Ceará, Minas Gerais (Map 11).
SPECIMENS EXAMINED. BAHIA: Mun. Abaíra, Gerais do Pastinho, 31 Jan. 1992, *Hind et al.* H 51405 (K); Mun. Caetité, Ramal a 29 km na estrada Caetité-Brumado, 19 Feb. 1992, *de Carvalho et al.* 3778 (K); São Francisco, 8 Feb. 1997, *Stannard et al.* PCD 5237 (K); Mun. Delfino, Serra do Curral Frio, 9 March

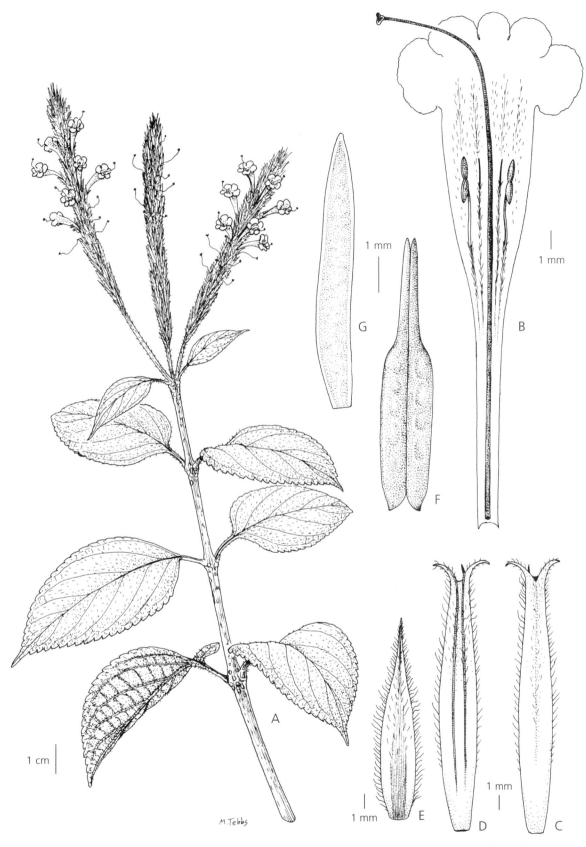

Fig. 14. *Stachytarpheta trispicata*. **A** habit; **B** corolla; **C** calyx adaxial view; **D** calyx abaxial view; **E** bract; **F** fruit; **G** fruit dissected along commissure. **A**, **B** from *Pereira* 9754; **C – E** from *Harley* 21692; **F**, **G** from *Harley* 21579. DRAWN BY MARGARET TEBBS.

1997, *Gasson et al.* PCD 6179 (K); Sento Sé, 29 April 1981, *Orlandi* 392 (K); Mun. Maracás, 26 km SW de Maracás, 27 April 1978, *Mori et al.* 9992 (K); 9 km de Maracás, 24 Jan. 1965, *Pereira & Pabst* 8589 (K). CEARÁ: Jaburuna, Planalto da Ibiapaba, 1 April 1995, *Bezerra* sn (K). MINAS GERAIS: Mun. Monte Azul, 7 km E of Monte Azul, 28 Jan. 1991, *Taylor et al.* 1465 (K).

HABITAT. *Caatinga*, especially where disturbed or degraded, and *campos gerais*.

CONSERVATION STATUS. Least concern. Reasonably common where found.

A treelet with bright red flowers. The inflorescence is thicker and the leaves less scabrid than in *Stachytarpheta scaberrima*.

Glaziou 11325 (K) collected from Espírito Santo appears to be *Stachytarpheta coccinea*, but the label data has "fl. bleues".

33. Stachytarpheta galactea *S. Atkins* sp. nov.

a ceteris speciebus sectionis statura suffruticis usque 50 cm alti (non fruticis alti vel arborae parvae), corollis albis (non rubris vel caeruleis) differt. Typus: Bahia, Mun. Caetité, São Francisco caminho para Lagoa Real, 8 Feb. 1997, *Saar et al.* PCD 5239 (holotypus ALCB).

Subshrub to 50 cm. Stem 4-sided, glabrous. Leaves patent, petiolate, chartaceous, ovate, 2.8 – 6 × 1.4 – 3 cm, apex acute, base truncate or attenuate, decurrent into petiole, margin crenate, upper surface with scattered uniseriate hairs, lower surface with network of prominent, reticulate nerves, all densely covered with uniseriate hairs. Inflorescence c. 6 × 1 cm; bracts narrowly triangular, c. 18 mm long, sparsely covered with long, uniseriate hairs. Calyx bifid, c. 18 mm long, outer surface with uniseriate hairs along veins. Corolla hypocrateriform, white, tube straight, c. 15 mm long. No further description possible from material available. Plate 6B.

Only known from the type.

HABITAT. Not known.

CONSERVATION STATUS. Data deficient.

This is a smaller plant than all the others in this group, and has pure white corollas. It does, however share the characters of a bifid calyx and a broad inflorescence. Unfortunately, although more material has been sought, none has been found. I am not able to include a drawing of this new species, but there is a photograph.

34. Stachytarpheta speciosa *Pohl ex Schauer* (1847: 568; 1851: 210, t. 35).

Type: 'In Brasilia medit. ad Salgado', *Pohl* s.n. (holotype W; isotypes K, 2 sheets both marked Herb. Imp. Vind. 1037 No. 172).

Shrub or subshrub, branched. Stems slightly rounded, covered with uniseriate hairs. Leaves chartaceous, elliptic, apex acuminate, base attenuate, decurrent into petiole, margin crenate-serrate, 3.5 – 7 × 1 – 2.5 cm, upper surface almost glabrous except for scattered hairs along veins, lower surface with uniseriate hairs, longer and more numerous along veins. Inflorescence medium length, straight, stout, 6 – 12 cm long × 10 mm wide; bracts subulate, 12 – 18 mm long, with scattered uniseriate hairs. Calyx c. 22 mm long, straight, bifid, lobes with 2 teeth, outer surface hairy, glabrous on inner surface. Corolla blue, infundibular, straight, tube 40 mm long, lobes c. 5 mm across, glabrous on outer surface, inner with hairs and glands. Stamens inserted towards top of corolla tube with anthers lying just below throat. Ovary and style glabrous, style 40 mm long, ovary 2 mm long. Fruit light brown, c. 10 mm long, slightly elongated at apex, with a prominent attachment scar.

DISTRIBUTION. Uncertain. Pohl's original specimen was collected reportedly from Minas Gerais. Most other collections are of cultivated material.

SPECIMENS EXAMINED. RIO DE JANEIRO: Itatiaia, 25 Aug. 1989, *Avila* 15 (K); SÃO PAULO: Cultivada nos viveiros da Prefeitura, 8 July 1937, *Etzel* s.n. (K); Cultivada no Jardim Botânico de São Paulo, 15 Feb. 1967, *Peteado* 1 (K); Reserva Biológica Parque Est. das Fontes do Ipiranga, 5 April 1983, *Ussui & Silva* 33 (K).

HABITAT. Apart from cultivated specimens, there are no modern collections or data for this species.

CONSERVATION STATUS. Critically endangered. It is uncertain whether or not this species still survives in the wild.

A remarkable plant with corolla tubes 4 cm long. Cannot be mistaken for any other species.

Group 5 Radlkoferiana

Eight species in Brazil.

Key to species in the Radlkoferiana group

2. Bracts obovate or ovate, conspicuous · 3
    Bracts oblong-ovate, angustate, elliptic, linear, or as leaves, not conspicuous · 5
3. Leaves sessile, erect, imbricate, elliptic to narrowly elliptic, margin of upper leaves deeply revolute · · · · · · ·
    · 40. **S. ganevii**
    Leaves petiolate, patent, rhombic, broadly spathulate or broadly elliptic, margin only slightly inrolled · · · · 4
4. Decumbent shrub; leaves broadly elliptic · 41. **S. arenaria**
    Erect shrub; leaves rhombic to broadly spathulate · 39. **S. almasensis**
5. Leaves sessile, imbricate, linear-elliptic to long-triangular, margin deeply inrolled · · · · · 35. **S. radlkoferiana**
    Leaves short-petiolate, not imbricate, ovate, obovate or subrotund, margin not deeply inrolled · · · · · · · · · 6
6. Leaves ovate, upper and lower surface densely covered in long white hairs, giving a silvery-grey
    appearance, margin not revolute · 36. **S. lychnitis**
    Leaves rotund or obovate, upper surface of leaves strigose or almost glabrous and glossy, lower surface · · · ·
    densely strigose or lanate, margin revolute · 7
7. Shrub to 1.5 m.; leaves rotund, 1.5 – 2.8 × 1.4 – 2.5 cm, undersurface lanate · · · · · · · · · · · · 37. **S. froesii**
    Shrub to 40 cm.; leaves obovate, 0.8 – 2.2 × 0.4 – 1.5 cm, undersurface strigose · · · · · · · · 38. **S. bromleyana**

---

35. Stachytarpheta radlkoferiana *Mansf.* in Pilg. (1924: 156); Harley & Simmons (1986); Zappi *et al.* (2003). Type: Brazil, Bahia, Itubira, Carrasco, 1700 m, 1914, *Luetzelburg* 214 (holotype B†; isotype fragment TEX). Isotypes may exist elsewhere.

Shrub to 2 m, branched. Stems woody, rounded, glabrous to densely white-pubescent, more pubescent at apex of plant, hairs simple. Leaves sessile, erect, patent or deflexed, imbricate; somewhat fleshy, linear-elliptic to long-triangular to narrow-oblong, 0.7 – 1.5 × 0.4 – 0.8 cm, apex obtuse,

base truncate to slightly auriculate, margin deeply revolute, sometimes covering lower surface, upper surface with shiny sessile glands, slightly wrinkled, glabrous or with tuft of white hairs at base, lower surface densely white-hairy or lanate. Inflorescence short, up to 2 cm long × 10 mm wide, rachis obscured by upper leaves with only corollas extended (flowers almost appear to be solitary in the axils of the upper leaves); bracts leaf-like, identical to upper leaves. Calyx 8 mm long, deeply bifid, lobes equal without teeth, outer surface pubescent. Corolla bright red, hypocrateriform, tube straight, c. 18 mm long, slightly enlarged at middle, inner surface with ring of long white hairs at throat, outer surface with glandular hairs, lobes 5, small, all ± equal, c. 2 mm across. Stamens inserted towards bottom of tube, included. Ovary 1.5 mm long; style long exserted, c. 15 mm long. Fruit black, surface reticulate with prominent attachment scar. Plate 4C.

Leaves erect, linear-elliptic to long-triangular, lower
    surface densely white-hairy · · a. var. **radlkoferiana**
Leaves patent to deflexed, narrow-oblong, lower
    surface lanate · · · · · · · · · · · · · · · · b. var. **lanata**

a. var. **radlkoferiana**

Leaves erect, imbricate; somewhat fleshy, linear-elliptic to long-triangular, 0.7 – 1.2 × 0.4 – 0.6 cm, apex obtuse, base truncate to slightly auriculate, margin deeply in-rolled, sometimes covering lower surface, upper surface with shiny sessile glands, slightly wrinkled, glabrous except for tuft of white hairs at base, lower surface densely white hairy. Fig. 15.

Map 12. Distribution of *Stachytarpheta radlkoferiana* var. *radlkoferiana* (●).

DISTRIBUTION. Bahia, mostly within the Chapada Diamantina National Park (Lençois – Mucugê) (Map 12).

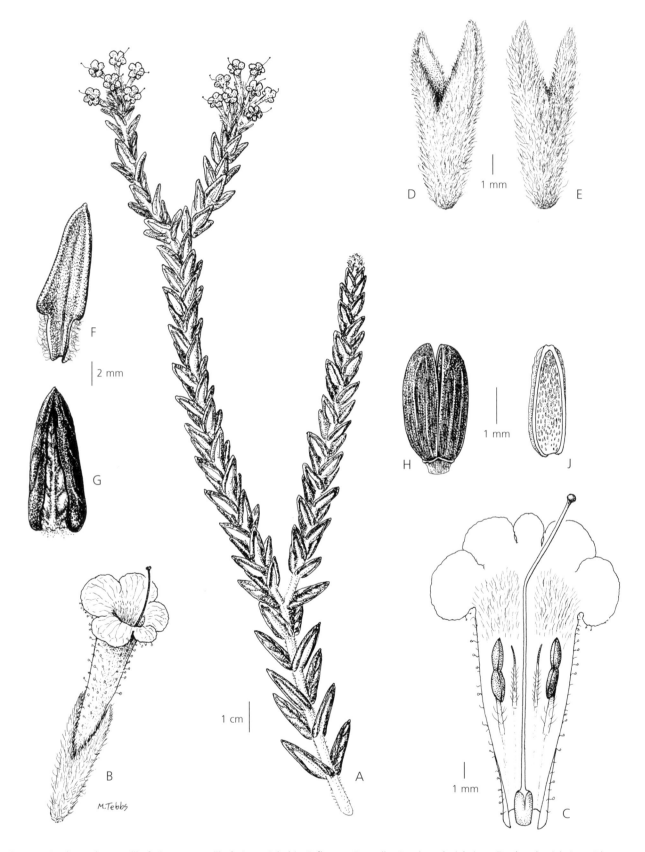

M.Tebbs

Fig. 15. *Stachytarpheta radlkoferiana* var. *radlkoferiana*. **A** habit; **B** flower; **C** corolla; **D** calyx adaxial view; **E** calyx abaxial view; **F** bract; **G** leaf; **H** fruit; **J** fruit dissected along commissure. **A, F, G** from *Hatschbach* 47946; **B – E, H, J** from *Harley et al.* 15398. DRAWN BY MARGARET TEBBS.

SPECIMENS EXAMINED. BAHIA: Mun. Mucugê: By Rio Cumbuca, 3 km S of Mucugê, 4 Feb. 1974, *Harley et al.* 15938 (K); Na estrada para Jussiape, 3 km S de Mucugê, 26 July 1979, *Mori et al.* 12604 (K); Serra do Sincorá, c. 15 km NW of Mucugê on road to Guiné, 26 March 1980, *Harley et al.* 20981 (K); Margem da estrada Mucugê – Cascavel, 20 July 1981, *Menezes et al.* SPF 18355/ CFCR 1443 (K); Rio Paraguaçu, 15 June 1984, *Hatschbach & Kummrow* 47946 (K). Mun. Piatã, arredores da cidade no caminho para a Capelinha, 4 Nov. 1987, *Harley et al.* 24173 (K). Mun. Andaraí, 20 km from Mucugê on road to Andaraí, 14 April 1990, *Carvalho & Thomas* 3055 (K). Mun. Abaíra – Catolés, Serra do Barbado, 26 Feb. 1994, *Atkins et al. CFCR* 14586 (K). Mun. Abaíra, caminho Capão – Bicota, 2 April 1992, *Ganev* 157 (K); Caminho Boa Vista – Bicota, 23 July 1994, *Ganev* 3421 (K); Riacho da Taquara, 28 Jan. 1992, *Stannard & Silva* H 50846 (K); Descida para Tijuquinho, 6 Jan. 1992, *Harley et al.* H 50649 (K); Campo de Ouro Fino, 25 Feb. 1992, *Læssøe & Sano* H 52335 (K); Serra ao Sul do Riacho da Taquara, 10 Jan. 1992, *Harley et al.* H 51291 (K); Serra do Bicota, 5 July 1993, *Ganev* 1804 (K).

HABITAT. Restricted to Bahia in damp sandy soils by rivers at higher altitudes (850 – 1500 m), along the eastern/northwards area of the Chapada.

CONSERVATION STATUS. Vulnerable and dependent on conservation. Several localities within Parque Nacional da Chapada Diamantina, and therefore protected.

An attractive plant with narrow, dark green leaves and bright red flowers.

b. var. **lanata** *S. Atkins* **var. nov.** a var. *radlkoferiana* foliis patentibus (non ascendentibus vel strictis) anguste oblongis (non lineato-lanceolatis) infra magis lanato-indumentosis praecipue in foliis superis, marginibus minus revolutis differt. Typus: Bahia, Alto do Morro do Pina, 20 July 1981, *Giulietti et al.* SPF 18463/CFCR 1551 (holotype SPF; isotype K).

Shrub to 1 m. Leaves sessile, patent to deflexed, narrow-oblong, 1 – 1.5 × 0.4 – 0.8 cm, upper surface glabrous, lower surface lanate, margin inrolled.

BAHIA. Serra do Sincorá, c. 15 km NW of Mucugê on road to Guiné, 26 March 1980, *Harley et al.* 20968 (K).

HABITAT. Found at lower altitudes and in drier conditions than *Stachytarpheta radlkoferiana* var. *radlkoferiana*.

CONSERVATION STATUS. Conservation dependent.

36. Stachytarpheta lychnitis *Mart. ex Schauer* (1847: 571; 1851: 217); Harley & Simmons (1986). Type 'in altis Serra das Lages et Sincorá, nec non in campis editis ad Villa do Rio de Contas', *Martius* s.n. (holotype M n.v.).

Subshrub to 50 cm, branched, sometimes with prostrate stems; woody rootstock. Stem woody, rounded, densely covered with white, upward-pointing hairs. Leaves erect to patent, thick-chartaceous, ovate, 1.4 – 3 × 0.7 – 1.5 cm, apex acute, base short-attenuate, margin crenate, outline obscured by hairs, upper and lower surface covered with dense white hairs. Inflorescence up to 3 cm long by up to 20 mm wide; bracts herbaceous, becoming woody, elliptic, covered with long white hairs especially along margin. Calyx c. 14 mm long, 2-toothed, densely covered in soft, white hairs. Corolla red, hypocrateriform, tube straight, up to 25 mm long, lobes small, c. 4 mm across. Stamens attached just below middle of tube. Ovary c. 2 mm long, style 20 mm long. Fruit dark brown, smooth. Fig. 16, Plate 6A.

DISTRIBUTION. Bahia: restricted to an area around Barra da Estiva, north to Mucugê, and south to Ituaçu (Map 13).

SPECIMENS EXAMINED. BAHIA: Mun. Barra da Estiva: Serra do Sincorá, c. 6 km N of Barra da Estiva on Ibicoara road, 28 Jan. 1974, *Harley et al.* 15533 (K); c. 14 km N of Barra da Estiva near the Ibicoara road, 2 Feb. 1974, *Harley et al.* 15853A (K); Barra da Estiva on the Capão da Volta road, 22 March 1980, *Harley et al.* 20702 (K); Estrada Ituaçu – Barra da Estiva, a 12 km de Barra da Estiva, próximo ao Morro do Ouro, 18 July 1981, *Giulietti et al.* SPF

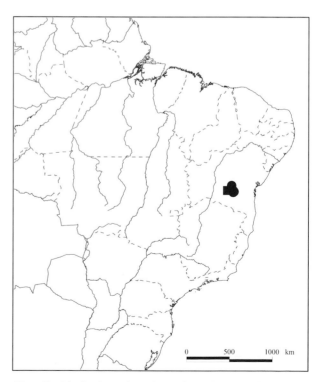

Map 13. Distributions of *Stachytarpheta almasensis* (■) and *S. lychnitis* (●).

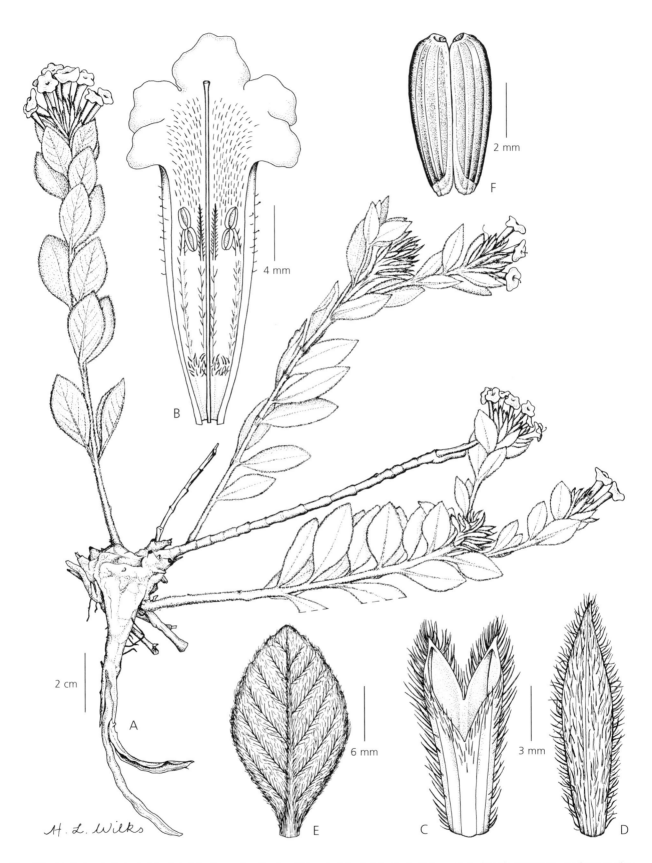

Fig. 16. *Stachytarpheta lychnitis.* **A** habit; **B** corolla; **C** calyx; **D** bract; **E** leaf; **F** fruit. **A, E** from *Hatschbach* 48359; **B – D** from *Furlan et al.* CFCR 2047; **F** from *Harley et al.* 15533. DRAWN BY HAZEL WILKS.

18103/CFCR 1240 (K); Capão da Volta, 19 Sept. 1984, *Hatschbach* 48359 (K); Morro do Ouro, 9 km ao S da cidade na estrada para Ituaçu, 16 Nov. 1988, *Harley et al.* 26467 (K); Barra da Estiva, camino a Ituaçu, Morro do Ouro e Morro da Torre, 22 Nov. 1992, *Arbo et al.* 5697 (K). Mun. Mucugê: Mucugê, estrada Mucugê – Guiné a 28 km de Mucugê, 7 Sept. 1981, *Furlan et al.* SPF 18842/CFCR 2047 (K); Estrada Barra da Estiva – Mucugê, km 7, 4 July 1983, *Coradin et al.* 6390 (K).

HABITAT. In open *campo rupestre*/grassland; c. 1100 – 1300 m. Eastern Chapada Diamantina.

CONSERVATION STATUS. Near threatened. Although restricted to a small area, a relatively common plant where it occurs.

A low shrub, it differs from the other species in the group by its grey-silvery leaves.

37. **Stachytarpheta froesii** *Moldenke* (1949: 173). Type: Bahia, Serra do Sincorá, 19 Feb. 1943, *R. de Lemos Froes* 20140 (holotype NY; isotype K).

Shrub to 1.5 m, much branched. Stem woody, somewhat gnarled, outer layer hairy, but often splitting to reveal lighter inner bark. Leaves short-petiolate, crowded at apex of stem, coriaceous, rotund, 1.5 – 2.8 × 1.4 – 2.5 cm, apex rounded, base cuneate, decurrent into petiole, margin crenate, revolute, upper surface glossy, rugose, older leaves hairy only along nerves, lower surface lanate, with fine simple hairs forming a mat. Inflorescence straight, short, c. 2 cm long × 10 mm wide (rachis obscured by upper leaves with only corollas extended); bracts linear, 4 mm long, densely lanate. Calyx straight, 9 mm long, densely lanate on outer surface, hairy within, 2-lobed. Corolla red, infundibular, tube straight, c. 11 mm long, lobes c. 2 mm across. Stamens attached just above middle, sessile. Ovary glabrous, c. 2 mm long, style 1.9 cm long. Fruit dark brown, surface warted, with prominent attachment scar.

DISTRIBUTION. Bahia: Pai Inácio (Map 14).

SPECIMENS EXAMINED: BAHIA: Mun. Palmeiras: Pai Inácio, 21 Nov. 1994, *Melo et al.* PCD 1177 (K); 25 Oct. 1994, *Carvalho et al.* PCD 983 (K); 24 May 1980, *Harley et al.* 22489 (K); 29 June 1983, *de Queiroz* 633 (K).

HABITAT. *Campo rupestre*, among rocks at 900 – 1200 m.

CONSERVATION STATUS. Endangered and conservation dependent. See under *Stachytarpheta tuberculata* (p. 197).

A large distinctive shrub with leathery leaves, lanate on the undersurface, and red flowers. This species has a very limited distribution, and has not been collected much since the type in 1943.

38. **Stachytarpheta bromleyana** *S. Atkins* **sp. nov.** a *S. lychnite* foliis obovatis usque cuneiformibus 0.8 – 2.2 × 0.4 – 1.5 cm (non ovatis, 1.4 – 3 × 0.7 – 1.5 cm), strigosis (non sericeis) distincta. Typus: Bahia, Serra do Sincorá, NW face of Serra de Ouro, 9 km S of Barra da Estiva, 1300 – 1500 m, 24 March 1980, *Harley et al.* 20861. (holotypus CEPEC; isotypus K).

Subshrub to 40 cm, much branched. Stems rounded, woody, gnarled, covered with a tangle of uniseriate hairs, more glabrous towards base of plant. Leaves petiolate, patent, thickly chartaceous, obovate to cuneiform, 0.8 – 2.2 × 0.4 – 1.5 cm, apex obtuse, base cuneate, margin crenate, slightly revolute, upper surface strigose, lower surface densely strigose. Inflorescence 1 – 2 cm long × 10 mm wide; bracts oblong-ovate, 5 – 6 mm long, strigose. Calyx 9 mm long, with 2 teeth, strigose on outer surface. Corolla bright red, infundibular, tube c. 15 mm long, hairy at throat, lobes c. 4 mm across. Stamens lying at middle. Ovary flask-shaped, c. 1.5 mm long, style c. 16 mm long. Fruit black, shiny, slightly reticulate, without prominent attachment scar. Fig. 17.

Known only from the type (Map 14).

HABITAT. On steep rocky north face at 1300 – 1500 m.

CONSERVATION STATUS. Data deficient. Only one collection of this species. This may be due to difficulty of access — a steep scree ascent. It does not appear to occur on neighbouring mountains which have been

Map 14. Distributions of *Stachytarpheta arenaria* (●), *S. bromleyana* (■) and *S. froesii* (▲).

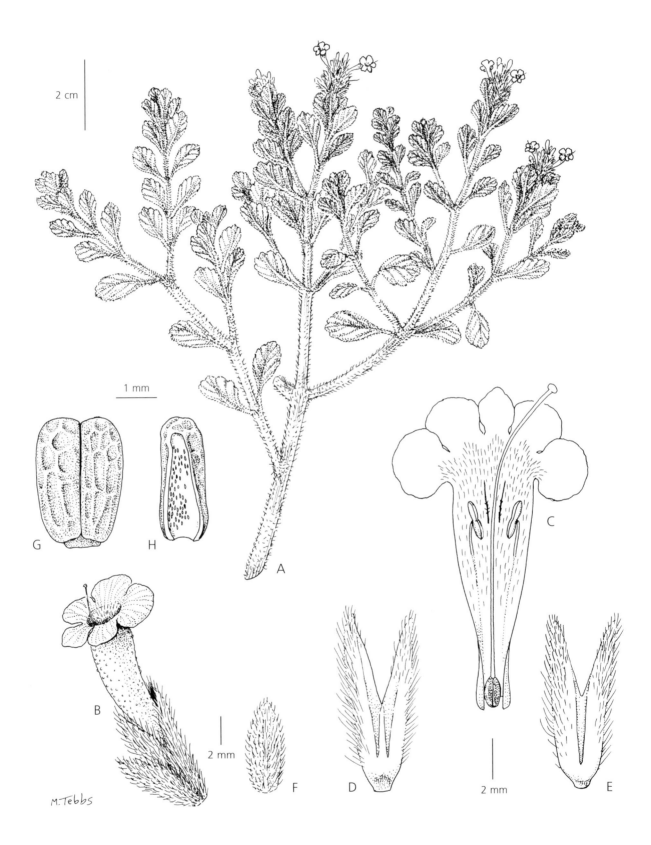

Fig. 17. *Stachytarpheta bromleyana.* **A** habit; **B** flower; **C** corolla; **D** calyx adaxial view; **E** calyx abaxial view; **F** bract; **G** fruit; **H** fruit dissected along commissure. All from *Harley et al.* 20861. DRAWN BY MARGARET TEBBS.

collected many times. The whole area may suffer from burning for agricultural use so the species may be under threat, even Critically Endangered.

This species is close to *Stachytarpheta arenaria*, but of an upright habit, and *S. lychnitis*, but has differently shaped leaves, and a strigose, not sericeous indumentum. This species is named for Mrs Gail Bromley, who first identified it as a new species.

39. Stachytarpheta almasensis *Mansf.* in Pilg. (1924: 155); Atkins in Stannard (1995: 628); Zappi *et al.* (2003). Type: Bahia, Serra das Almas, 1600 m, 1914, *Luetzelburg* 253 (holotype B†; isotype M).

Subshrub to 80 cm, much branched. Stems woody, rounded, covered in rigid white uniseriate hairs. Leaves short-petiolate, patent, herbaceous, rhombic to broadly spathulate, (5 –)10 – 20(– 30) × (3 –)8 – 16(–20) mm, apex obtuse, base narrowly cuneate, margin entire at base, broadly crenate above, revolute, both surfaces covered with short, white bristly uniseriate hairs, more numerous below. Inflorescence 1.5 – 4 cm long × 15 – 20 mm wide; bracts herbaceous, membranous, ovate to broadly ovate, tinged pink, c. 7 mm long. Calyx c. 12 mm long, 2-lobed, lobes without teeth, outer surface covered in glandular hairs. Corolla bright red, hypocrateriform, tube straight, c. 20 mm long, lobes small, ± equal. Stamens inserted at middle. Ovary ovate, c. 2 mm long, style c. 20 mm long. Fruit dark brown, shiny, with reticulate surface.

DISTRIBUTION. Bahia: Pico das Almas (Map 13).
SPECIMENS EXAMINED. BAHIA. Mun. Rio de Contas: Pico das Almas, vale ao sudeste do Camp do Queiroz, 1 Dec. 1988, *Harley et al.* 26559 (K); Pico das Almas, valley SE of Campo do Queiroz, 2 Dec. 1988, *Fothergill* 67 (K).
HABITAT. Restricted to *campo rupestre* on the Pico das Almas at 1400 – 1500 m.
CONSERVATION STATUS. Vulnerable. The Pico das Almas is within a Municipal Park, but there is now some disturbance caused by ecotourism and agricultural use. Even here, it is not a common species, and could be under threat.

Inflorescence similar to *Stachytarpheta ganevii*, with broad, pinkish bracts, but the leaves are spathulate, not narrowly triangular.

40. Stachytarpheta ganevii *S. Atkins* **sp. nov.** a ceteris speciebus sectionis inflorescentiis conspicuis usque 6 cm longis et foliis imbricatis margine valde incurvatis differt. Typus: Bahia, Mun. de Rio de Contas, Gerais do Porco Gordo, 16 July 1993, *Ganev* 1869 (holotypus HUEFS; isotypus K).

Shrub/subshrub to 1 m from woody rootstock. Stems rounded, covered with white, uniseriate hairs. Leaves sessile, erect, imbricate, coriaceous, elliptic to narrowly elliptic, 8 – 18 × 2.5 – 7 mm, apex acute, base truncate, margin deeply inrolled, upper surface with scattered uniseriate hairs, lower surface covered with sessile glands, with long uniseriate hairs along main vein. Inflorescence 1.5 – 6 cm long × 20 – 25 mm wide, bracts ovate to broadly obovate, apex apiculate, 9 mm long. Calyx 12 mm long, basically 2-lobed, outer surface glabrous, glandular, inner surface glabrous. Corolla red (pinky-red), hypocrateriform, tube enlarged just above middle, c. 16 mm long, densely white-hairy at throat, lobes c. 3 mm across. Stamens attached in lower part of tube, anthers lying at middle. Ovary ovoid, c. 1 mm long, style 20 mm long. Fruit dark brown, reticulate with prominent attachment scar. Fig. 18, Plate 3A.

DISTRIBUTION. Bahia: Mun. Abaíra and Mun. Rio de Contas (Map 15).
SPECIMENS EXAMINED. BAHIA: Mun. Abaíra, Riacho das Taquaras, 1700 m, 21 May 1992, *Ganev* 333 (HUEFS; K); Campo de Ouro Fino, 1650 m, 12 Feb. 1992, *Nic Lughadha et al.* H. 52011 (K); 16 July 1992, *Ganev* 668 (HUEFS; K); Campo do Cigano, 1700 – 1800 m, 29 Feb. 1992, *Nic Lughadha et al.* H. 52396 (K)
HABITAT. *Campo rupestre*, among rocks, at altitudes between 1100 – 1700 m.
CONSERVATION STATUS. Vulnerable. Areas could be subject to burning, and together with a restricted distribution, this species could be under threat.

A quite spectacular plant with somewhat leathery, dark green leaves, and a prominent, bracteate inflorescence, with bracts and calyces tinged red and a pink corolla. The inflorescence is similar to that of *Stachytarpheta almasensis*, and the leaves are similar to those of *S. radlkoferiana*. This species is named for Wilson Ganev who first collected it.

41. Stachytarpheta arenaria *S. Atkins* **sp. nov.** ab aliis speciebus sectionis habitu decumbenti non erecto; a *S. ganevii* foliis ovatis patentibus (non ellipticis imbricatisque) distinguenda. Typus: Bahia, Mun. de Rio de Contas, Serra dos Brejões, 1400 m, 9 Aug. 1993, *Ganev* 2061 (holotypus HUEFS; isotypus K).

Decumbent shrub to 70 cm, much branched. Stems rounded, covered in uniseriate hairs. Leaves short-petiolate, patent, coriaceous, broadly elliptic, 7 – 11 × 5 – 9 mm, apex obtuse, base attenuate, margin inrolled, upper surface sparsely to densely covered with uniseriate hairs, lower surface densely covered with uniseriate hairs. Inflorescence 2 – 3 cm long by 15 mm wide; bracts obovate, 6 – 7 mm long, apex apiculate. Calyx 10 mm long, 2-lobed, outer surface

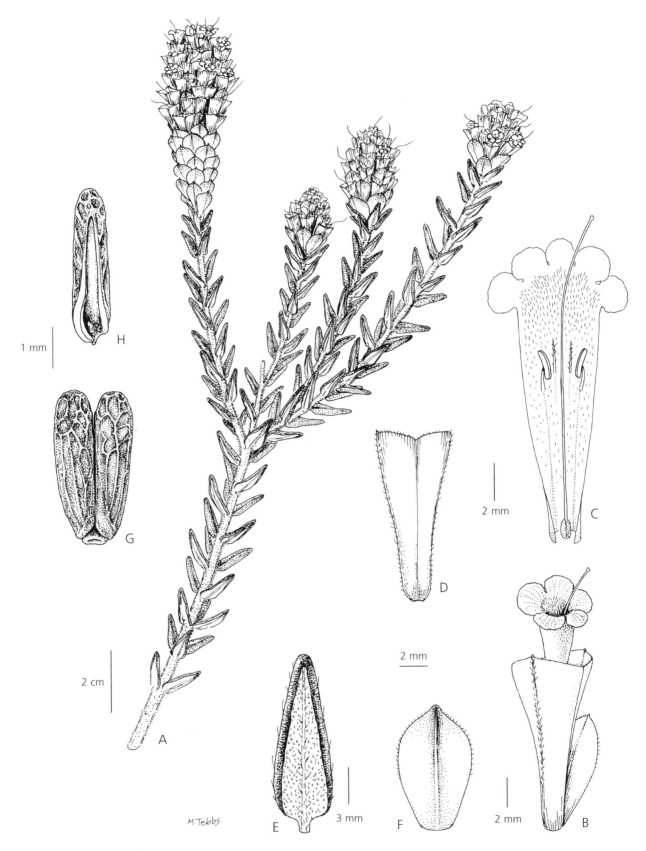

Fig. 18. *Stachytarpheta ganevii*. **A** habit; **B** flower; **C** corolla; **D** calyx; **E** leaf undersurface; **F** bract; **G** fruit; **H** fruit dissected along commissure. **A, D, G, H** from *Ganev* 333; **B, C, E, F** from *Ganev* 1869. DRAWN BY MARGARET TEBBS.

Map 15. Distributions of *Stachytarpheta ganevii* (●) and *S. piranii* (■).

glabrous, with sessile glands and very short hairs at apex, glabrous on inner surface. Corolla red, hypocrateriform, tube not enlarged, c. 16 mm long, densely white-hairy on inner surface at throat, lobes c. 1 mm across. Stamens attached just below middle of corolla tube, anthers lying at middle. Ovary ovoid, 1 mm long; style 15 mm long. Fruit dark brown, reticulate surface. Fig. 19A – E.

DISTRIBUTION. Bahia: Mun. Abaíra, Mun. Rio de Contas (Map 14).
SPECIMENS EXAMINED. BAHIA: Mun. Abaíra, Serra dos Cristais, 1960 m, 8 June 1994, *Ganev* 3331 (HUEFS; K); Serra das Brenhas, 1860 m, 22 Oct. 1992, *Ganev* 1305 (HUEFS; K). Mun. Rio de Contas, Fazendola, 16 Nov. 1996, *Bautista et al.* PCD 4329 (ALCB).
HABITAT. In *campo rupestre*, in sandy soil at altitudes between 1400 and 1960 m.
CONSERVATION STATUS. Vulnerable. These areas are not protected, and are now subject to pressures from ecotourism and encroaching agriculture. Could be under threat.

This is a decumbent shrub with a similar inflorescence to *Stachytarpheta almasensis* and *S. ganevii*, but the leaves are different from both.

42. Stachytarpheta piranii *S. Atkins* **sp. nov.** a ceteris speciebus sectionis habitu arborescenti/arbusculae et foliis magnis glabris differt. Typus: Bahia, Mucugê, Margem da estrada Andaraí-Mucugê, a 13 km de Mucugê, 21 July 1981, *Pirani et al.* SPF 18532 (holotypus SPF, isotypus K).

Treelet to 4 m. Stem 4-sided at base, rounded above, grey, covered with very short, white, upward-pointing hairs. Leaves petiolate, patent, coriaceous, ovate, 3.8 – 13.5 × 1.8 – 6.5 cm, apex obtuse, base attenuate, margin almost entire, barely crenate towards apex, upper and lower surface glabrous. Inflorescence 4.5 – 6 cm long by c. 10 mm wide; bracts obovate, c. 12 mm long, striate, glabrous. Calyx 2-lobed, 20 mm long, glabrous. Corolla bright red, tube straight, c. 36 mm long. Stamens lying half-way up tube. Ovary flask-shaped, c. 2 mm long. Fruit dark brown, reticulate, style attached below apex, with prominent attachment scar, c. 5 mm long. Fig. 19F – K.

DISTRIBUTION. Only two specimens known, both from Bahia, Mun. Mucugê (Map 15).
SPECIMENS EXAMINED. BAHIA: Mucugê, Chapada Diamantina, 20 Sept. 1998, *Guedes et al.* 6167 (ALCB).
HABITAT. *Campo rupestre*.
CONSERVATION STATUS. Critically endangered. Only 2 collections appear to have been made, in spite of several attempts to recollect. The area around the type locality has been cleared of natural vegetation and planted with coffee, even though it is within the Chapada Diamantina National Park.

The label information calls this a treelet of 4 m. The large, showy bracts make the inflorescence close to that of *Stachytarpheta ganevii*, but it differs very markedly in habit, and in its large glabrous leaves. This species is named for José R. Pirani who first collected it.

Group 6 Villosa
Ten species in Brazil.

Key to species in the Villosa group

1. Leaves completely glabrous ·· 2
   Leaves not glabrous ··· 5
2. Stems reed-like, striate, easily flattened; internodes long, 5 – 13.5 cm long ············ 50. **S. integrifolia**
   Stems not reed-like, woody, resistant; internodes shorter, 3 – 8 cm long ···················· 3

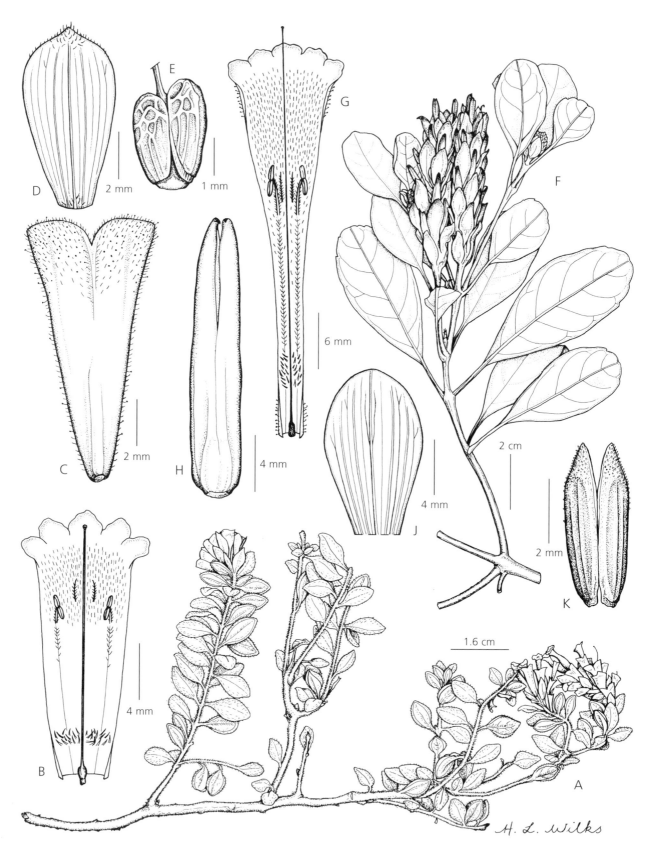

Fig. 19. *Stachytarpheta arenaria*. **A** habit; **B** corolla; **C** calyx; **D** bract; **E** fruit. *Stachytarpheta piranii*. **F** habit; **G** corolla; **H** calyx in bud; **J** bract; **K** fruit. **A – E** from *Ganev* 2061; **F – K** from *Pirani et al.* CFCR 1660/SPF 18532. DRAWN BY HAZEL WILKS.

3. Leaves 2 at each node; ovate to broadly ovate to subrotund; bracts broadly triangular, shortly attenuate
at apex, very short, c. 3 mm long, barely visible · 4
Leaves often 3 at each node, obovate or rhombic; bracts narrowly triangular, long-attenuate at apex,
c. 5 – 7 mm long, almost half as long as calyx · 49. **S. rhomboidalis**
4. Leaves ovate, remotely crenate in upper part of leaf; inflorescence nodding · · · · · · · · · · · · · 51. **S. glauca**
Leaves ovate to broadly ovate to subrotund, entire; inflorescence erect · · · · · · · · · · · · · 52. **S. atriflora**
5. Shrub forming clumps up to 2 m across; inflorescence to 33 (– 73) cm long; flowers densely to
sparsely spread along inflorescence axis · 45. **S. longispicata**
Shrub not clump-forming; inflorescence to 9 cm long; flowers densely packed along inflorescence axis · · · · 6
6. Flowers red or orange-red · 43. **S. villosa**
Flowers dark purple to black · 7
7. Bract and calyx finely covered with hairs, but not so much as to obscure the surface and nerves · · · · · · · · ·
· 44 . **S. puberula**
Bract and calyx densely covered with hairs so as to obscure the surface and nerves from view · · · · · · · · · · 8
8. Indumentum on calyx and bracts lanate · 46. **S. glazioviana**
Indumentum on calyx and bracts sericeous · 9
9 Leaves elliptic to broadly elliptic, 1 – 3.5 × 0.6 – 1.5 cm · 48. **S. sericea**
Leaves ovate to broadly ovate to subrotund, 2.5 – 8.5 × 1.9 – 5 cm · · · · · · · · · · · · · · · · 47. **S. dawsonii**

---

43. Stachytarpheta villosa (*Pohl*) *Cham.* (1832: 247). As to type excluding specimen cited. Type: Goyaz (Goiás), ad Cavalcante, September 1819, *Pohl* s.n. (holotype W n.v.; isotype BR).

Melasanthus villosus Pohl (1827: 76, pl. 60). Type as above.

S. goyazensis Turcz. (1863: 198) Type: Goiás, March 1840, *Gardner* 3936 (isotypes K, G, NY).

S. schaueri Moldenke (1941a: 22) superfl. nom. nov. for *S. villosa*.

S. mollis Moldenke (1947b: 370). Type: Goyaz (Goiás), entre Sobradinho et Lagoa do Mestre d'Armas, Nov. – Dec. ?1895, *Glaziou* 21906 (holotype S; isotype G).

Subshrub to 50 cm with woody rootstock, much-branched. Stems rounded, densely white-hairy to densely villous. Leaves sometimes with smaller leaves in the axils, subcoriaceous, ovate, apex obtuse to rounded, base cuneate, decurrent into petiole, (2 –)3 – 3.5(– 4) × (1 –)2.5 – 2.8 cm, upper and lower surface densely covered in white uniseriate hairs. Inflorescence straight, erect, 3.5 – 4.5(– 9) cm long by c. 2.5 – 3 cm wide; bracts broadly triangular, c. 8 – 10 mm long, hairy. Calyx ± equally 5-lobed, straight, c. 14 – 16.5 mm long, fairly densely covered with white uniseriate hairs. Corolla red, or orange-red, hypocrateriform, tube straight, c. 15 – 16.5 mm long, lobes c. 2.5 mm wide, throat glandular, ring of hairs at base of tube above ovary. Stamens lying halfway down tube. Ovary oblong. Fruit dark brown, c. 3 mm long, with prominent attachment scar. Fig. 20.

DISTRIBUTION. Distrito Federal and Goiás.
SPECIMENS EXAMINED. DISTRITO FEDERAL: c. 15 km E of Brasília, 17 Aug. 1965, *Irwin et al.* 7801 (K); Reserva

Águas Emendadas, 24 Jan. 1978, *Krapovickas & Cristobal* 33188 (K); Bacia do Rio São Bartolomeu, 16 March 1981, *Heringer et al.* 6433 (K); Estrada de Brasília para Unaí, 15 Feb. 1986, *de Carvalho & Fráguas* 2285 (K). **GOIÁS:** Chapada dos Veadeiros, 14 km S of Veadeiros, 25 April 1956, *Yale Dawson* 14659 (RSA); 65 km due N of Brasília, 22 Dec. 1968, *Harley & Barroso*, 11470 (K); 4 km S of Teresina, 18 March 1973, *Anderson* 7410 (K); Norte de São João da Aliança, 12 Feb. 1990, *Hatschbach & Nicolack* 53839 (K); Alrededores de Colinas, 5 Feb. 1990, *Arbo et al.* 3704 (K); 1828 – 1830, *Burchell* 8029 (K).

HABITAT. *Cerrado* at 900 – 1100 m.

CONSERVATION STATUS. Least concern. A fairly common plant in the *cerrados* of Goiás and Distrito Federal. Seems to survive in grazed areas.

A species with wide variation in size of leaves and bracts and in the indumentum. *Harley & Barroso* 11470 has smaller, less hairy leaves; *Hatschbach et al.* 70422 has smaller but hairier leaves, and longer, more pointed bracts; *Irwin et al.* 7801 has more pointed bracts.

Pohl first used the specific epithet *villosa* in 1827, but under the generic name of *Melasanthus*. Chamisso in 1832 then misapplied the name under *Stachytarpheta*; even though he assumed he was describing the same taxon, the Sellow specimen he was considering was in fact *S. commutata* Schauer. However, according to Art. 7.4 of the ICBN (Saint Louis Code, Greuter *et al.* 2000), 'A new name formed from a previously published legitimate name is, in all circumstances, typified by the type of the basionym, even though it may have been applied erroneously to a taxon now considered not to include that type'. The correct name for this taxon, therefore is *Stachytarpheta villosa* (Pohl) Cham.

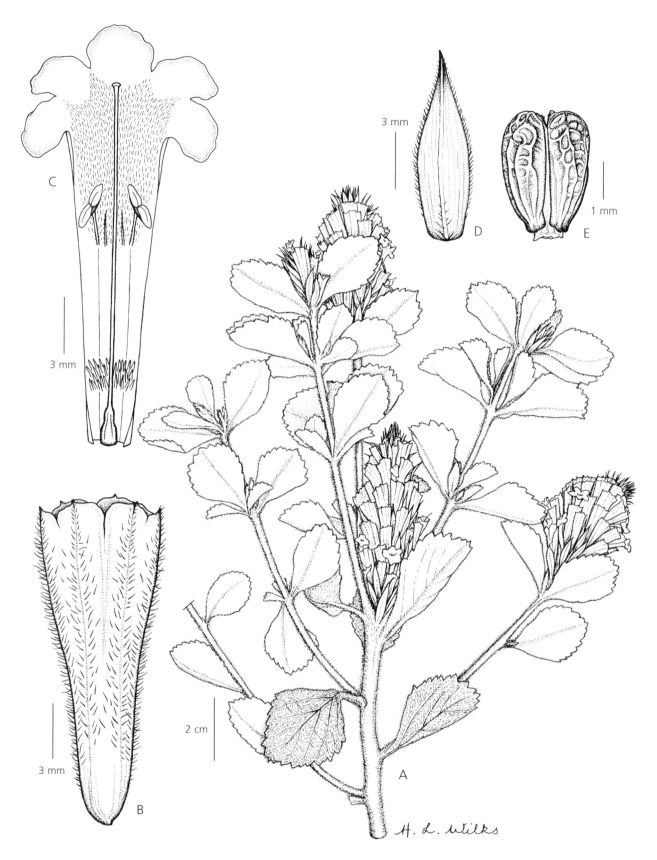

Fig. 20. *Stachytarpheta villosa.* **A** habit; **B** calyx; **C** corolla; **D** bract; **E** fruit. **A – D** from *França et al.* 4622; **E** from *Anderson* 7410.
DRAWN BY HAZEL WILKS.

Pohl described the flower as "atropurpurea". Although there is some variation of flower colour, it is dark red to dark orange. There are no collections which I have seen, nor in the field, of a similar plant with dark purple flowers. I did make notes on some collections that the calyx is dark purple. It is a fairly widespread, common species with quite a lot of variation.

44. Stachytarpheta puberula (*Moldenke*) *S. Atkins* **stat. nov.**

Stachytarpheta rhomboidalis var. *puberula* Moldenke (1979b: 473). Type: Goiás, Alto Paraíso, Chapada dos Veadeiros, 5 km E of Alto Paraíso, 26 Jan. 1979, *Gates & Estabrook* 69 (holotype MICH; fragment NY).

Shrub to 1 m, branched. Stem rounded, puberulent. Leaves sessile, erect, thickly chartaceous, ovate, 1.5 – 2.5 × 0.9 – 1.5 cm, apex obtuse, base attenuate, margin entire in lower half, obscurely crenate in upper half, upper and lower surface sparsely puberulous. Inflorescence erect, to 2.5 cm long by 2 cm wide; bracts triangular, c. 5 mm long. Calyx ± equally 5-lobed, straight, c. 12 mm long, very sparsely puberulous. Corolla dark purple, hypocrateriform, tube straight, c. 15 mm long.

Known only from the type.
HABITAT. Sandy, rocky hillside at 1500 m.
CONSERVATION STATUS. Data deficient. Only 1 collection of this species, even though the area has been the subject of many expeditions. The area is now under the protection of the Chapada dos Veadeiros National Park.

This species is similar to *Stachytarpheta villosa*, but has a dark purple corolla, and is puberulous, not villous.

45. Stachytarpheta longispicata (*Pohl*) *S. Atkins* **comb. nov.**

Melasanthus longespicatus Pohl (1827: 77, tab. 61). Type: Brazil, 'Habitat in montosis aridis, ad Serra de Cristaes Capitaniae Goyaz [Goiás]', *Pohl* s.n. (holotype W n.v.; possible isotype BR).
S. chamissonis Walp. (1844: 10) nom. illeg. Type: as for *Melasanthus longespicatus* Cham.

Shrub to 2 m with woody rootstock, branched. Stem rounded, sparsely to densely covered in uniseriate hairs. Leaves petiolate, patent, sometimes with smaller leaves in the axils, thickly chartaceous, spathulate, fan-shaped, ovate or sub-rotund, 0.5 – 5.5 × 0.3 – 3 cm, apex obtuse to rounded, base truncate or attenuate or cuneate, margin crenate, upper surface rugose, with reticulate nerves slightly impressed, lower surface with prominent reticulate nerves, whole surface covered in uniseriate hairs, more numerous along nerves. Inflorescence long, broad, slightly curving, 8 – 73 × 2 – 2.5 cm, rachis visible between flowers, flowers sessile or petiolate; bracts linear or narrowly triangular with shoulders, somewhat woody, c. 3 – 7 mm long, densely covered in uniseriate hairs. Calyx a narrow tube, tapering towards base, c. 10 – 16 mm long, densely covered in uniseriate hairs, more dense at base, 5-toothed. Corolla is variously described as salmon pink, orange, red, dirty red, pale red, rust (Pohl calls it dark red), hypocrateriform, tube straight, c. 18 mm long, lobes small, hardly spreading, c. 3 mm across. Stamens inserted at middle of tube. Ovary ovoid, c. 1 mm long; style 20 mm long. Fruit dark brown, 5 mm long, without prominent attachment scar.

There are many different forms of this taxon. The typical subspecies is based on Pohl's original description and plate, and on specimens collected from the same area as the type specimen, i.e., Serra dos Cristais. However, this is not the best known form; the one usually associated with the name *Stachytarpheta chamissonis* is now assigned to *S. longispicata* subsp. *ratteri*, which has bigger leaves and a longer inflorescence, and is common around Brasília.
See Table 4.

a. subsp. longispicata

var. longispicata

Clump-forming shrub (up to 2 m across) to 1 m with woody rootstock, branched. Leaves petiolate, patent, sometimes with smaller leaves in the axils, thick-chartaceous, fan-shaped, 1.5 – 4 × 1.2 – 3 cm, apex obtuse to rounded, base truncate or cuneate. Inflorescence long, broad, slightly curving, 25 – 33 × 2 – 2.5 cm, rachis visible between flowers; bracts narrowly triangular with shoulders, somewhat woody, c. 6 mm long, densely covered in uniseriate hairs. Calyx c. 12 mm long. Corolla variously described as orange, red, dirty red, pale red, rust (Pohl calls it dark red), hypocrateriform, tube straight, c. 18 mm long, lobes small, hardly spreading, c. 3 mm across. Plate 7A, B.

DISTRIBUTION. Goiás (Map 16).
SPECIMENS EXAMINED. GOIÁS: Serra dos Cristais, 9 km S of Cristalina on road to Catalão, 4 April 1973, *Anderson* 8104 (K); Mun. de Cristalina, 4 Feb. 1987, *Pirani et al.* 1613 (K); Cristalina, 21 Feb. 1992, *Mello-Silva et al.* 559 (K); Mun. Niquelândia, 15 km N of Niquelândia, 21 April 1988, *Brooks et al.* 157 (K); 1 km após a Mina da Companhia de Níquel Tocantins, 12 May 1996, *Mendonça et al.* 2428 (K).

Table 4. Showing differences between the infraspecific taxa of *Stachytarpheta longispicata*

Taxon	Leaf shape	Leaf size (cm)	Inflorescence length (cm)	Bract (mm)	Calyx length (mm)	Corolla tube length (mm)	Geography	Plant size (m)
S. longispicata subsp. *longispicata*	fan-shaped	1.5 – 4 × 1.2 – 3	25 – 33	6	12	18	Goiás, Cristalina	1
var. *andersonii*	ovate	2 – 4 × 0.8 – 1.5	15	6	13	16	Goiás, North, near border with Tocantins	2
var. *longipedicellata*	ovate	2.5 – 5.5 × 1 – 2.5	17	7	12	16	Goiás, Alto Paraíso, Chapada dos Veadeiros	2
var. *parvifolia*	ovate	1.5 – 3 × 0.9 – 1.4	7 – 9	5	10	11	Goiás, Chapada dos Veadeiros	0.5
subsp. *ratteri*	fan-shaped	3 – 5 × 2.5 – 4	25 – 73	7	16	20	Distrito Federal	2
subsp. *brevibracteata*	ovate to subrotund	1.2 – 2.5 × 0.5 – 1.2	7 – 8	3	12	16	Minas Gerais, Morro das Pedras	1
subsp. *minasensis*	spathulate to ovate	0.5 – 2 × 0.3 – 1	4 – 7	4	8	10	Minas Gerais, Serra do Cabral	0.35

HABITAT. *Cerrado* and *campo rupestre* up to 1200 m; often on nickel-rich substrates.
CONSERVATION STATUS. Least concern.

var. **andersonii** (*Moldenke*) *S. Atkins*, **comb. nov.**
S. chamissonis var. *andersonii* Moldenke (1974b: 467). Type: Brazil, Goiás, Serra Geral do Paraná, 4 km by road E of São João da Aliança, 24 May 1973, *Anderson et al.* 7893 (holotype TEX; isotype NY).

Shrub to 2 m, branched. Leaves ovate, 2 – 4 × 0.8 – 1.5 cm, base attenuate, decurrent into petiole; inflorescence to 15 cm, flowers pedicellate; bracts linear to 6 mm; calyx to 13 mm; calyx covered with short uniseriate hairs, patent or erect; corolla red-orange.

DISTRIBUTION. North-eastern Goiás.
SPECIMENS EXAMINED. GOIÁS: Mun. Água Fria de Goiás, 8 May 2000, *Hatschbach et al.* 70677.
HABITAT. *Campo rupestre*/rocky *cerrado*.
CONSERVATION STATUS. Data Deficient. Neither of these localities is in a protected area.

This taxon has ovate leaves, not fan-shaped, and a shorter inflorescence than the typical variety.

var. **longipedicellata** (*Moldenke*) *S. Atkins* **comb. nov.**
S. chamissonis var. *longipedicellata* Moldenke (1974b: 467). Type: Goiás, Chapada dos Veadeiros, 6 March 1973, *Anderson et al.* 6460 (holotype TEX; isotype NY).
S. chamissonis var. *longipetiolata* Moldenke (1983a: 414). Type: Goiás, Chapada dos Veadeiros, 24 March 1971, *Irwin et al.* 33117 (isotypes NY, K).

Shrub to 2 m, branched. Leaves ovate, 2.5 – 5.5 × 1 – 2.5 cm, base attenuate, decurrent into petiole; inflorescence to 17 cm, flowers pedicellate, bracts linear to 7 mm. Calyx to 12 mm densely covered with uniseriate hairs pointing in all directions; corolla salmon pink.

DISTRIBUTION. Goiás: Chapada dos Veadeiros (Map 16).
SPECIMENS EXAMINED. GOIÁS: Chapada dos Veadeiros, 6 km NE of Alto Paraíso, 14 Feb. 1979, *Matiko Sano & Filgueiras* 47 (K); 30 km from Alto Paraíso, 30 May 1994, *Ratter et al.* 7271 (K).
HABITAT. Hilltop *cerrado*. 1250 – 1600 m.
CONSERVATION STATUS. Data deficient. Only three collections, but all from within the protected area of the Parque Nacional da Chapada dos Veadeiros.

Similar to var. *andersonii*, but with salmon pink corolla.

var. **parvifolia** (*Moldenke*) *S. Atkins*, **comb. nov.**
S. chamissonis var. *parvifolia* Moldenke (1980: 39). Type: Goiás, Chapada dos Veadeiros, 14 Feb. 1979, *Gates & Estabrook* 176 (isotype NY).

Shrub to 50 cm, branched. Leaves ovate 1.5 – 3 × 0.9 – 1.4 cm; bracts narrowly triangular to 5 mm; calyx to 10 mm; calyx and rachis densely covered with uniseriate hairs pointing in all directions; corolla red.

Known only from the type.
HABITAT. Rocky outcrops at 1500 m.
CONSERVATION STATUS. Data Deficient. Only 1 collection from within the protected Parque Nacional de Chapada dos Veadeiros.

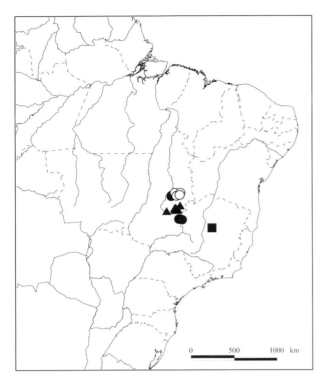

Map 16. Distributions of *Stachytarpheta longispicata* subsp. *longispicata* (●), *S. longispicata* subsp. *minasensis* (■), *S. longispicata* subsp. *ratteri* (▲) and *S. longispicata* var. *longipedicellata* (○).

b. subsp. ratteri *S. Atkins* **subsp. nov.** a subsp. *longispicato* foliis majoribus (5 × 4 cm non 2 – 4 × 1.5 – 2.5 cm), inflorescentiis densioribus, calyce intense viridi vel flavo, corolla aurantiaca vel chryso-brunnea vel 'terracotta' differt. Typus: Distrito Federal, Fazenda Água Limpa near Vargem Bonita, 15 March 1976, *Ratter & Fonsêca* 2775 (holotypus K).

Clump-forming shrub with woody rootstock to 2 m, unbranched. Leaves petiolate, 3 – 5 × 2.5 – 4 cm. Inflorescence up to 73 cm long, crowded especially towards apex, where rachis is not visible; bracts linear, c. 7 mm long. Calyx c. 16 mm long; corolla hypocrateriform, dull orange, brownish orange or terracotta, tube c. 20 mm, lobes c. 2 mm across. Fig. 21A – C.

DISTRIBUTION. Distrito Federal, especially around Brasília (Map 16).
SPECIMENS EXAMINED. DISTRITO FEDERAL: Lagoa do Paranoá, 11 April 1968, *Philcox & Onishi*, 4767 (K); Chapada da Contagem, 6 km para Brasilândia, 5 Feb. 1987, *Pirani et al.* 1650 (K); Parque Nacional de Brasília, 22 Jan. 1978, *Krapovickas et al.* 33180 (K); 4 Feb. 1992, *Barros et al.* 2220 (K).
HABITAT. *Cerrado*, usually disturbed, at c. 1000 m.
CONSERVATION STATUS. Vulnerable. The rapidly expanding population of Brasília threatens the

environment of Distrito Federal (Filgueiras 1997). This means encroachment of protected areas for housing, habitat alteration caused by weeds, and human-provoked fires. This variety has a limited distribution and could be at threat.

This variety is what most people associate with the name *Stachytarpheta chamissonis* (now called *S. longispicata*). However, Pohl based his description on specimens collected around Cristália, and with a shorter inflorescence; subsp. *ratteri* is confined to an area within Distrito Federal. This subspecies is named for James Ratter who first collected it.

c. subsp. brevibracteata *(Moldenke) S. Atkins*, **comb. & stat. nov.**
S. chamissonis var. *brevibracteata* Moldenke (1980: 38). Type: Minas Gerais, Morro das Pedras, 25 km NE of Patrocínio, 28 Jan. 1970, *Irwin et al.* 25457 (type NY).

Subshrub to 1 m, branched. Leaves ovate to subrotund, 1.2 – 2.5 × 0.5 – 1.2 cm, base attenuate, decurrent into petiole; inflorescence to 8 cm, flowers pedicellate; bracts linear to 3 mm. Calyx to 12 mm, covered in short indumentum of uniseriate hairs; corolla dark red.

Known only from the type.
HABITAT. Sandy and gravelly *campos* and *cerrado* outcrops. 1050 m.
CONSERVATION STATUS. Data deficient. Only one collection. Area not protected.

Smaller leaves and shorter inflorescence than type subspecies.

d. subsp. minasensis *S. Atkins* **subsp. nov.** a subsp. *longispicata* foliis minoribus 0.5 – 2 × 0.3 – 1 cm (non 1.5 – 4 × 1.2 – 3 cm), inflorescentia usque 7 cm (non usque 33 cm), bracteis linearibus usque 4 mm (non triangularibus usque 6 mm) distincta. Typus: Minas Gerais, Serra do Cabral, Mun. Joaquim Felício, 15 April 1996, *Hatschbach et al.* 64789 (holotypus MBM; isotypus K).

Shrub to 35 cm with woody rootstock, branched. Leaves spathulate to ovate, 0.5 – 2. × 0.3 – 1 cm; bracts linear to 4 mm; calyx to 1 cm; calyx and rachis with uniseriate mostly patent hairs.

DISTRIBUTION. Minas Gerais: Serra do Cabral (Map 16).
SPECIMENS EXAMINED. MINAS GERAIS: Mun. Joaquim Felício, Serra do Cabral, 15 May 2001, *Hatschbach et al.* 72033 (K).
CONSERVATION STATUS. Data deficient. Only two collections from the Serra do Cabral. The area is not

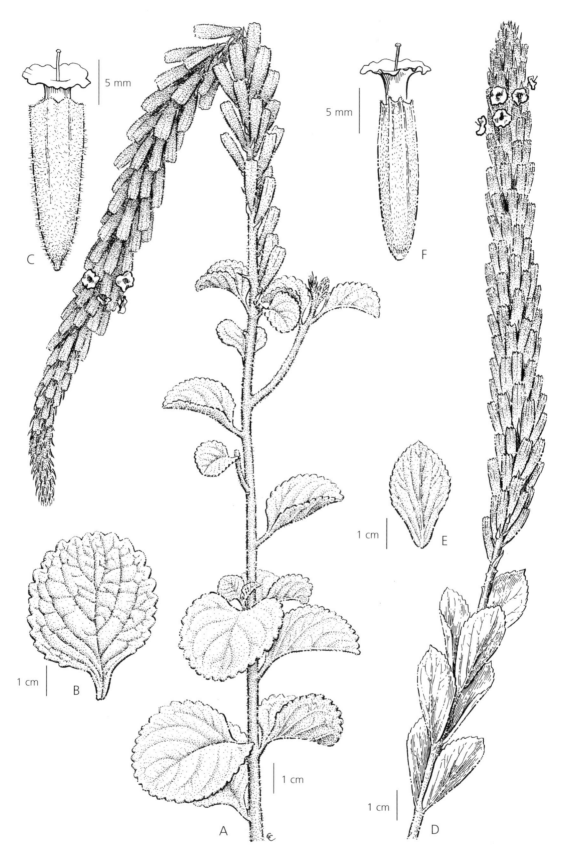

Fig. 21. *Stachytarpheta longispicata* subsp. *ratteri*. **A** inflorescence; **B** leaf; **C** flower. *Stachytarpheta sericea* × *S. longispicata* subsp. *ratteri*. **D** inflorescence; **E** leaf; **F** flower. **A – C** from *Ratter & da Fonsêca* 2775; **D – F** from *Pirani et al.* 1524. DRAWN BY ELEANOR CATHERINE.

protected and likely to suffer a decline in habitat quality in the future due to local extractive activities (Taylor & Zappi 2004).

The leaves are very small. The taxon is known only from a small area.

46. Stachytarpheta glazioviana *S. Atkins* **sp. nov.** a *S. sericea* foliis late ovatis apice obtusis usque rotundatis (non ellipticis usque late ellipticis apice acutis usque apiculatis) indumento lanato (non sericeo) distincta. Typus: Goiás: Serra de São Joaquim, au Morro do Salto, *Glaziou* 21905 (holotypus P; isotypus K).

S. glazioviana Loes. ex Glaz. (1911: 544) nom. nud.

Subshrub, not branched. Stems rounded, covered with long white simple hairs, lying in all directions, but appressed to the stem. Leaves sessile, erect, decussate, coriaceous, broadly ovate, 2.8 – 4 × 1.8 – 3.2 cm, apex obtuse to rounded, base truncate to slightly cordate, margin crenate, upper and lower surface densely covered in long white, soft simple hairs. Inflorescence 2 – 4 × c. 2 cm; bracts narrowly triangular, c. 7 mm long, densely lanate. Calyx c. 10 mm long, densely lanate with soft simple hairs. Corolla black, infundibular, straight, tube c. 12 mm long, lobes c. 1 mm across. Stamens lying half way up tube. Ovary asymmetrical c. 1 mm long. Fruit not seen.

DISTRIBUTION. Goiás.
SPECIMENS EXAMINED. GOIÁS: Entre Sobradinho et Lagoa do Mestre d'Armas, 1896, *Glaziou* 21907 (K).
HABITAT. Not known, possibly *cerrado.*
CONSERVATION STATUS. Data deficient. It is not possible to pinpoint exactly where these specimens were collected (Wurdack 1970). There are no modern collections and it is to be supposed that this taxon is at risk.

The name *Stachytarpheta glazioviana* Loes. was first published in "Liste des Plantes du Brésil en 1861 – 1895" (Glaziou 1911). It has been shown (Atkins 1991) that the names were never validated, either by Glaziou or Loesener.

47. Stachytarpheta dawsonii *Moldenke* (1956: 231); Dawson (1957). Type: Goiás, 5 km W of Veadeiros, 29 April 1956, *Yale Dawson* 14722 (holotype RB n.v.; fragment TEX).

Subshrub to 1 m, unbranched, or branching just below inflorescence. Stems rounded, densely covered in white uniseriate hairs, mostly upward pointing, more dense towards top of plant. Leaves sessile to petiolate, patent or erect, subcoriaceous, ovate to

broadly ovate to subrotund, 2.5 – 8.5 × 1.9 – 5 cm, apex rounded, base attenuate, decurrent into petiole, margin serrate, upper and lower surface densely covered in white uniseriate hairs. Inflorescence 2 – 7 × c. 3 cm, bracts ovate with subulate apex, 4 mm long, densely sericeous. Calyx c. 16 mm long, densely sericeous, with 5 ± equal teeth. Corolla almost black, hypocrateriform, tube straight, c. 18 mm long, hairy and glandular at throat, lobes c. 2 mm across. Stamens lying just below throat. Ovary slightly asymmetrical, c. 2 mm long, style 19 mm long. Fruit dark brown to black, 5 mm long.

DISTRIBUTION. Goiás (Map 17).
SPECIMENS EXAMINED. GOIÁS: *Campo* between São Domingos and Posse, May 1840, *Gardner* 4339; Rio da Prata, c. 6 km S of Posse, 6 April 1966, *Irwin et al.* 14455 (NY).
HABITAT. Dry, upland *cerrado* at 800 m.
CONSERVATION STATUS. Near threatened. I have looked for this species without success. There is still *cerrado* between São Domingos and Posse, but also agricultural encroachment. This may be at risk.

Moldenke's original description says that the bracts are lanceolate, 10 – 15 mm long. I have not seen the type and the fragment of the type from TEX has no bracts. There is a photograph of the specimen published in a report of the Machris Brazilian Expedition (Dawson 1957). This species differs from

Map 17. Distributions of *Stachytarpheta atriflora* (●) and *S. dawsonii* (■).

Stachytarpheta glazioviana and *S. sericea* in having much larger leaves, and from *S. glazioviana* because the leaves are sericeous, not lanate, and from *S. sericea* by its less imbricate leaves. There is one specimen, *Collares & Fernandez* 134, collected in Bahia, at São Desidério, on the road to Sítio Grande, in open tree savana. If this locality is correct, then the distribution extends to north-western Bahia.

48. Stachytarpheta sericea *S. Atkins* (1991: 281); Dawson (1957). Type: Goiás, entre Engenho & Jatobá, dans le campos, 21 Jan. 1895, *Glaziou* 21904 (holotype P; isotype K).
S. sericea Loes.; Glaz. (1911: 544) *nom. nud.*

Shrub to 1 m, unbranched, or branching just below inflorescence. Stems 4-sided, densely covered with white hairs. Leaves sessile, erect, decussate, coriaceous, elliptic to broadly elliptic, 1 – 3.5 × 0.6 – 1.5 cm, apex acute to apiculate, base truncate, margin serrate, upper and lower surface densely covered with white hairs. Inflorescence up to 8.5 × 2 cm; bracts narrowly triangular, glabrous, 6 mm long. Calyx 14 mm long, entire, outer surface densely sericeous, inner surface less densely hairy. Corolla dark purple-black, hypocrateriform, tube straight, c. 15 mm long, densely hairy at throat, lobes c. 4 mm across. Stamens inserted at middle, included. Ovary flask-shaped, c. 2 mm, style c. 20 mm long. Fruit black, surface reticulate. Fig. 22, Plate 4A.

DISTRIBUTION. Goiás.
SPECIMENS EXAMINED. GOIÁS: entre St Luzia & Engenho, 13 Sept. 1895, *Glaziou* 21903 (K); Cristalina area, 5 km da cidade, estrada para Paracatu, 4 Feb. 1987, *Pirani et al.* 1520 (K); c. 3 km N, 2 March 1966, *Irwin et al.* 13232 (K); c. 7 km by road NW of Cristalina on road to Brasília, 3 April 1973, *Anderson* 8058 (K); Cristalina, Serra do Cristal, 21 Feb. 1975, *Hatschbach et al.* 36393 (K); Cristalina, 29 Jan. 1980, *Heringer & Rizz* 17581(K); Cristalina, 27 March 1963, *Pereira* 7348 (K); Cristalina, 2 Feb. 1967, *Heringer* 11338 (K); Serra Geral do Paraná, 7 km by road S of São João da Aliança, 22 March 1973, *Anderson* 7672 (K); c. 10 km S of São João da Aliança, 17 March 1971, *Irwin et al.* 32004 (K).
HABITAT. Planalto and high altitude *campo rupestre* and *cerrado*, often on quartzite; 950 – 1250 m.
CONSERVATION STATUS. Least concern. This species is still reasonably common within its range and is thought not to be at risk.

Stachytarpheta sericea appears to hybridise with *S. longispicata*, with which it grows in proximity. The hybrids can take the leaf form and habit from *S. longispicata* and the inflorescence shape, and flower colour from *S. sericea* or vice versa. The pollen of a putative hybrid had reduced or absent protoplasm (Atkins 1991):

Stachytarpheta sericea × longispicata. Fig. 21D – F.

GOIÁS: Contagem, 15 km de São Domingos, 15 May 2000, *Hatschbach et al.* 71139 (K); Alto Paraíso, rod. para Colinas do Sul, 14 June 1993, *Hatschbach et al.* 59512 (K); Cristalina, estrada para Paracatu, 4 Feb. 1987, *Pirani et al.* 1524 (K).

49. Stachytarpheta rhomboidalis (*Pohl*) *Walp.* 1845: 10); Schauer (1847: 569; 1851: 212).
Melasanthus rhomboidalis Pohl (1828: 78, t. 62). Type: 'Habitat in Montosis altis, locis aridis, ad S. Antonio de Montes Claros dicta, ante Meyaponte Capitaniae Goyaz (Goiás)', Jan. 1819, *Pohl* s.n. (holotype W n.v.; isotype BR).
Stachytarpheta triphylla (Pohl) Walp. (1845: 10).
Melasanthus triphyllus Pohl (1828: 79, t. 63). Type: "Habitat in Montosis siccis aridis, ad S. Antonio de Montes Claros, Captaniae Goyaz (Goias)", Jan. 1819, *Pohl* s.n. (holotype W n.v.; isotype BR).

Woody herb to 60 cm unbranched. Stem 4-sided, covered with very short white, uniseriate, bristly hairs. Leaves sessile, erect, often in 3s, chartaceous, obovate or rhombic, 3 – 5 × 2 – 3 cm, apex obtuse to acute, base truncate, margin remotely serrate, glabrous. Inflorescence 7 – 11 × 2 – 2.5 cm. Bracts narrowly triangular with long attenuation at apex, 5 – 7 mm with scattered hairs and glands. Calyx straight, 12 mm long, glabrous but glandular on outer surface. Corolla black, hypocrateriform, tube slightly constricted below throat, 15 mm long, glabrous on outer surface, very hairy at throat, lobes c. 2 mm across. Stamens inserted in middle part of tube. Ovary flask-shaped, c. 2 mm long, fruit not seen.

DISTRIBUTION. Goiás (Map 18).
SPECIMENS EXAMINED. GOIÁS: Chapada dos Veadeiros, Porto Seguro, 1896, *Glaziou* 21908 (G; K); entre Anápolis & Corumbá de Goiás, 2 April 1958, *Lima* 58-3008 (K); Estrada de Sto. Antonio do Rio Descoberto para Cidade Eclética, 29 Nov. 1965, *Cobra & Sucre* 401 (K).
HABITAT. In dry mountainous regions.
CONSERVATION STATUS. Vulnerable. There are no modern collections of this species. One locality, in the Chapada dos Veadeiros, is dubious (Wurdack 1970). The other localities, to the west of Brasília could well be endangered. The few collections, all from unprotected areas, make this a potentially vulnerable species.

This is another species from Goiás with black flowers. Can be differentiated from previous species by its glabrous leaves.

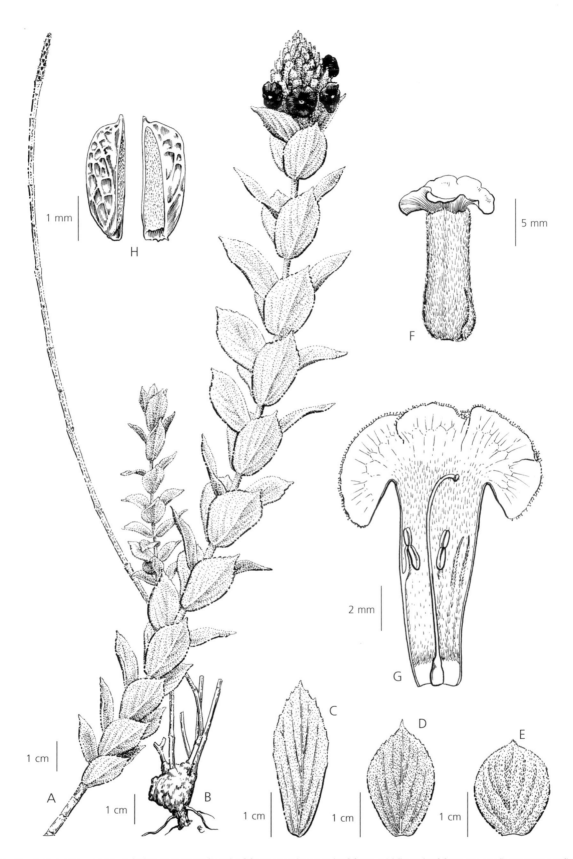

Fig. 22. *Stachytarpheta sericea.* **A** habit; **B** rootstock; **C** leaf from near base; **D** leaf from middle; **E** leaf from near inflorescence; **F** flower; **G** corolla; **H** fruit. **A – B** from *Anderson* 7672; **C – D**, **F** & **G** from *Pirani et al.* 1520; **E** from *Glaziou* 21904; **H** from *Glaziou* 21903.
DRAWN BY ELEANOR CATHERINE.

Map 18. Distributions of *Stachytarpheta integrifolia* (●) and *S. rhomboidalis* (■).

50. Stachytarpheta integrifolia (*Pohl*) *Walp.* (1845: 10).

Melasanthus integrifolius Pohl (1825: 80, t. 64). Type: 'in campis graminosis, ad Pillar, Capitaniae Goyaz (Goiás)', March & June 1819, *Pohl* s.n. (holotype W n.v.; isotype BR).

S. glauca var. *subintegrifolia* Schauer (1847: 569). Type: as above (holotype W n.v.; isotype BR).

Shrub to 1.6 m, unbranched. Stem reed-like, pithy, hollow, glabrous. Leaves sessile, erect or patent, widely spread with long internodes, thickly chartaceous, ovate, oblong or (subrotund), (leaves at apex of plant narrowly oblong), 3.5 – 6.5(– 11) × 1 – 4(– 9) cm, apex acute to obtuse, occasionally emarginate, base truncate or auriculate, sometimes amplexicaul, margin entire to coarsely serrate in upper part of leaf, upper and lower surface glabrous, tightly covered with small sessile glands, interspersed with larger flat open glands. Inflorescence 3 – 10 × 2.5 cm, somewhat nodding, bracts 7 mm long, triangular with scarious margin. Calyx straight, c. 20 mm long, regularly 5-toothed, teeth slightly rounded, topped with point, outer surface glabrous, but with one or two large flat glands. Corolla black, narrowly infundibular, 20 mm long, lobes c. 2 mm across. Stamens inserted at middle part of tube, more or less sessile. Ovary oblong, c. 2 mm long. Fruit dark brown, 7 mm long.

DISTRIBUTION. Goiás, around Niquelândia (Map 18).

SPECIMENS EXAMINED. GOIÁS: No locality, no date, Herb. Imp. Vien. no. 1907 (K); Serra Dourada, east of Formoso, 22 May 1956, *Yale Dawson* 15053 (RSA); Chapada dos Veadeiros, 4 Feb. 1990, *Arbo et al.* 3649 (K); leste de Niquelândia, 21 Jan. 1992, *Hatschbach & Kummrow* 56284 (K); Mun. Alto Paraíso, 16 Oct. 1990, *Hatschbach & Silva* 54662 (K); Mun. de Niquelândia, 7 May 1998, *Aparecida da Silva et al.* 3792 (K).

HABITAT. *Campo rupestre* on quartz; 485 – 1000 m.

CONSERVATION STATUS. Vulnerable. Some of these localities fall outside the protection of the Parque Nacional de Chapada dos Veadeiros and this species could be vulnerable.

Schauer (1847) reduced this species to a variety of *Stachytarpheta glauca*. He made Pohl's *Melasanthus glauca* into *S. glauca* var. *serratifolia* and *M. integrifolius* into var. *subintegrifolia*. There are enough differences between the species, I feel, to maintain *S. integrifolius*. It has a reed-like stem which is completely different from the stem of *S. glauca*, and long internodes; it has different shaped leaves and longer bracts. It is known from the modern collections of this species that it is confined to the area around Niquelândia. Unfortunately, at this stage, it is not known where *S. glauca* can be found.

51. Stachytarpheta glauca (*Pohl*) *Walp.* (1845: 11); Dawson (1957).

Melasanthus glaucus Pohl (1825: 81, t. 65). Type: 'in altis montosis, locis gramineis, in Serra S. Felis prope Rio Custodio' July & August 1819, *Pohl* s.n. (holotype W n.v.; istoype BR).

S. glauca var. *serratifolia* Schauer (1847: 569). Type: as above.

Shrub to 1 m, branched below inflorescence. Stem woody, rounded below, 4-sided above, glabrous. Leaves sessile, erect, chartaceous, ovate, 4 – 7 × 2 – 3.8 cm, apex acute or obtuse, base truncate, margin distantly serrate in upper part of leaf, upper and lower surface glabrous. Inflorescence 3.5 – 6 cm long by 20 mm wide, nodding; bracts triangular, c. 3.5 mm long, glabrous. Calyx with 4 regular teeth abaxially and shallow sinus adaxially, c. 12 mm long, glabrous. Corolla black with dark purple tube, hypocrateriform, 13 mm long, lobes c. 1 mm across. Stamens inserted at middle part of tube. Ovary oblong, c. 2 mm long. Fruit dark brown, c. 5 mm long.

DISTRIBUTION. Goiás.

SPECIMENS EXAMINED. Goiás, no locality, no date, Herb. Imp. Wien no. 1908 (K).

HABITAT. High, grassy areas.

CONSERVATION STATUS. Data deficient. No modern collections.

This species is very similar to *Stachytarpheta integrifolia* and *S. rhomboidalis*. It differs from *S. integrifolia* in not having a reed-like stem, and not having long internodes. It differs from *S. rhomboidalis* in having opposite leaves, and in having a nodding inflorescence. All have black corollas.

52. Stachytarpheta atriflora S. *Atkins* **sp. nov.** a *S. glauca* calyce longiore c. 15 mm longo (non 12 mm), bracteis brevioribus c. 2 mm longis (non 4 mm), venatione foliorum praecipue infra prominenti pallidaque (non tantum moderata concolorique), inflorescentia erecta (non nutanti) distinguenda. Typus: Goiás, Chapada dos Veadeiros, 8 Feb. 1987, *Pirani et al.* 1818 (holotypus SPF, isotypus K).

Woody herb or shrub to 1 m. Stems woody, rounded, glabrous. Leaves sessile, erect, coriaceous, ovate to broadly ovate or sub-rotund, 3 – 7 × 2 – 4 cm, apex rounded to obtuse, base truncate, margin entire to shallowly serrate in upper part of leaf on some specimens, upper and lower surface glabrous, tightly covered with small sessile glands, interspersed with larger flat open glands. Inflorescence 3 – 7 × 2 – 2.5 cm; bracts broadly triangular with shortly attenuate apex, c. 3 mm long. Calyx straight, 15 mm long, 4-toothed, teeth shallow with 2 shallow sinuses on each side, outer surface glabrous, but with 2 – 3 large flat glands like those on the leaves, towards the apex. Corolla black, narrowly infundibular, c. 20 mm long. Stamens

inserted at middle part of tube. Ovary narrowly-flask shaped, 3 mm long. Fruit pale yellow to light brown, 8 mm long. Plate 5C.

DISTRIBUTION. Goiás: Chapada dos Veadeiros (Map 17).
SPECIMENS EXAMINED. GOIÁS: Chapada dos Veadeiros, 13 March 1969, *Irwin et al.* 24277 (K); 23 March 1969, *Irwin et al.* 24941 (G); 24 March 1971, *Irwin et al.* 33116 (K); 8 March 1973, *Anderson* 6661 (K); 30 May 1994, *Ratter et al.* 7291 (K); 29 May 1994, *Ratter et al.* 7267 (K).
HABITAT. *Cerrado/campo rupestre* at 1000 – 1700 m.
CONSERVATION STATUS. Vulnerable. Conservation Dependent. All specimens are from within the Parque Nacional da Chapada dos Veadeiros, which is a strongly protected area.

Many of these specimens had been identified as *Stachytarpheta rhomboidalis*. However, they differ from *S. rhomboidalis* by the shape and size of the leaves, and in the leaves being only in opposite pairs, not in groups of three. Also the leaf margin is almost always entire, not serrate. This species is similar to *S. integrifolia* and *S. glauca*. It differs from *S. glauca* in its longer calyx, shorter bracts, its prominent and pale venation especially on the underside of the leaf, and in its erect inflorescence. It differs from *S. integrifolia*, by its stem not being reed-like, by the shape and size of the leaves, and by the short internodes.

Group 7 Procumbens
Four species in Brazil.

Key to species in the Procumbens group

1. Plants erect · 2
   Plants prostrate or decumbent · 3
2. Corolla tube up to 11 mm long · 55. **S. confertifolia**
   Corolla tube up to 30 mm long · 56. **S. monachinoi**
3. Bract as long or longer than calyx at anthesis; calyx and bract covered with long white hairs; flower white · 54. **S. candida**
   Bract half as long as calyx at anthesis; calyx and bract glabrous; flower pale blue/pale lilac · · 53. **S. procumbens**

53. Stachytarpheta procumbens *Moldenke* (1950b: 311). Type: Minas Gerais, Mun. Jaboticatubas, Serra da Ponte de Pedra, *Geraldo Mendes Magalhães* 2629 (holotype NY).

Decumbent subshrub with stems to 41 cm long, unbranched. Stems woody, rounded, fairly densely covered with white uniseriate hairs. Leaves sessile, subopposite, patent, chartaceous, oblong to obovate, 1.2 – 2 × 0.6 – 1 cm, apex obtuse to rounded, base truncate to rounded, margin entire or sometimes obscurely toothed towards apex,

finely ciliate, upper and lower surface almost glabrous with occasional large glands towards base of lamina. Inflorescence 2 – 5 cm long by 15 mm wide; bracts linear, c. 6 mm long, glabrous except for very small, fine hairs along margin. Calyx 10 mm long, with 4 ± equal teeth, outer surface glabrous. Corolla pale lilac, infundibular, tube straight, c. 10 mm long, inner surface with short glandular hairs at throat, lobes c. 5 mm across. Stamens inserted in upper part of tube. Ovary oblong, c. 1 mm long. Fruit light brown, surface reticulate, without stylopodium. Fig. 23.

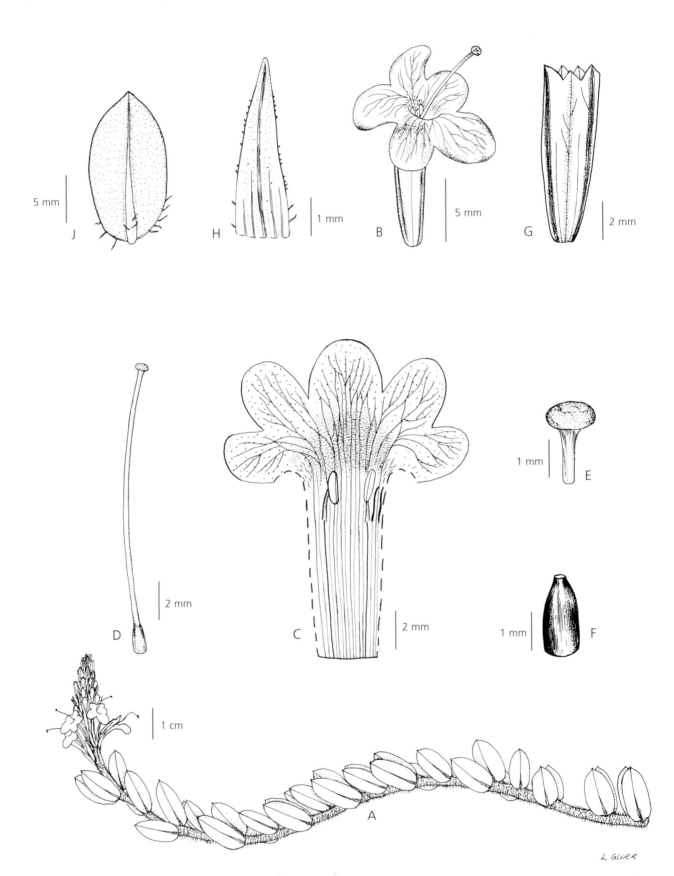

Fig. 23. *Stachytarpheta procumbens.* **A** habit; **B** flower; **C** corolla; **D** gynoecium; **E** stigma; **F** ovary; **G** calyx; **H** bract; **J** leaf undersurface. All from *Pirani et al.* CFCR 7488. DRAWN BY LINDA GURR.

DISTRIBUTION. Minas Gerais (Map 19).

SPECIMENS EXAMINED. MINAS GERAIS: Serra do Cipó, 6 April 1972, *Hatschbach* 29961 (NY); Mun. Santana do Riacho, Belo Horizonte-Conceição do Mato Dentro, 6 Oct. 1975, *Menezes* CFCR 7299/SPF 20124 (K); 5 Oct. 1981, *Furlan et al.* CFCR 7488/SPF 20295 (K); 30 Oct. 1981, *Henrique & Kawasaki* CFCR 7640/SPF 20430 (K).

HABITAT. *Campo rupestre*, grassland at edge of woodland.

CONSERVATION STATUS. Vulnerable. Localities within the Parque Nacional da Serra do Cipo are protected, those outside are not.

This species and the next are the only truly prostrate taxa in the genus. Both have woody rootstocks, and long ground-hugging stems, with only the inflorescence erect. *Stachytarpheta procumbens* has smaller leaves, and pale blue or lilac flowers, not white as in *S. candida*.

54. Stachytarpheta candida *Moldenke* (1968: 487). Type: Brazil, Goiás, Chapada dos Veadeiros, *Irwin et al.* 12393 (holotype US).

S. prostrata Loes. ex Glaz. (1911: 544) nom. nud.

S. candida f. *lilacina* Moldenke (1976: 374). Type: Goiás, Chapada dos Veadeiros, Mun. Alto Paraíso, 28 Sept. 1975, *Hatschbach & Kummrow* 37248 (holotype TEX).

Decumbent or prostrate shrub with stems to 24 cm, branching at base and sometimes below inflorescence. Stems rounded, covered in long white hairs pointing

Map 19. Distributions of *Stachytarpheta candida* (▲), *S. confertifolia* (●), *S. monachinoi* (■) and *S. procumbens* (○).

in all directions. Leaves sessile, patent, thickly chartaceous, elliptic, 2.5 – 5 × 0.8 – 1.5 cm, apex acuminate, base cuneate, margin entire from base to just above middle, serrate above, upper surface with sparse white hairs, but ± glabrous otherwise, lower surface more hairy with reticulate network of veins. Inflorescence 9 – 17 cm long × 10 mm wide; bracts narrowly triangular, c. 12 mm long, base truncate, apex narrowly attenuate, central keel covered in longish white hairs. Calyx straight, c. 12 mm long, covered in longish white hairs, 5-toothed. Corolla white, infundibular, tube c. 12 mm long, lobes c. 3 mm across with glandular hairs on outer surface and with long white hairs at throat. Stamens inserted about half way down tube. Ovary pear-shaped, c. 2 mm long. Fruit dark brown, surface reticulate, without stylopodium. Plate 2A.

DISTRIBUTION. Goiás (Map 19).

SPECIMENS EXAMINED. GOIÁS: 1896, *Glaziou* 21893 (K); Chapada dos Veadeiros, 9 Feb. 1966, *Irwin et al.* 12421 (NY); 21 Dec. 1968 (K), *Harley et al.* 11362 (K); 8 Feb. 1987, *Pirani et al.* 1859 (K); 9 Feb. 1987, *Pirani et al.* 1901 (K).

HABITAT. In open *campo* and *cerrado*.

CONSERVATION STATUS. Vulnerable. This is not a very widespread species, and although the localities are given as Chapada dos Veadeiros, they are not within the National Park boundary. This species could be vulnerable to the encroachment of agriculture into the *cerrado* vegetation.

See notes under previous species.

55. Stachytarpheta confertifolia *Moldenke* (1947c: 234). Type: Minas Gerais, Mun. Nova Lima, in a campo at Serra da Mutuca, 15 April 1945, *Williams & Assis*, 6639 (holotype GH).

S. chapadensis Moldenke (1973b: 117). Type: Brazil, Goiás, *Irwin, Harley & Smith* 32857 (holotype US n.v.; isotypes K, NY).

S. confertifolia var. *puberulenta* Moldenke (1975: 373). Type: Brazil, Goiás, Chapada dos Veadeiros, 20 Feb. 1975, *Hatschbach et al.* 36361 (holotype TEX; isotype NY).

Herb to 25 cm, unbranched. Stem rounded, covered with simple, upward-pointing hairs. Leaves sessile, erect, imbricate, sub-coriaceous, elliptic, 8 – 10 × 4 – 5 mm, apex acute, base truncate, entire, upper and lower surface glabrous. Inflorescence 2 – 3.5 cm long by 12 mm wide, bracts herbaceous, subulate, 8 mm long. Calyx 11 mm long, 4-toothed, teeth equal, shallow sinus, glabrous on outer surface, minutely hairy on inner surface. Corolla very pale lavender to white, hypocrateriform, hairy at throat, tube straight,

c. 11 mm long, lobes c. 4 mm across. Stamens lying just below throat, attached above middle of tube. Ovary narrowly flask-shaped, c. 2 mm long. Fruit pale, c. 5 mm long, without stylopodium.

DISTRIBUTION. Goiás and Minas Gerais (Map 19).
SPECIMENS EXAMINED. GOIÁS: Between Alto Paraíso de Goiás and Teresina de Goiás, 30 May 1994, *Ratter et al.* 7278 (K).
HABITAT. *Cerrado* and damp *campo limpo*, at 1250 – 1650 m.
CONSERVATION STATUS. Data deficient. This has a very restricted distribution and very few specimens have been collected, and even though localities within the Parque Nacional da Chapada dos Veadeiros are protected, this species may be at risk.

The type specimen has fine short white hairs along the stem, and the bracts are narrower. However, it is a good match for *Hatschbach* 36361, which was collected in the same area as the other specimens.

56. Stachytarpheta monachinoi *Moldenke* (1958: 328). Type: Brazil, Minas Gerais, Mun. Buenópolis, Serra do Cabral, 13 Dec. 1953, *Mendes Magalhães* 6024 (holotype TEX).

Herb with woody rootstock to 30 cm, unbranched. Stem rounded, glabrous to very sparsely covered with short hairs. Leaves sessile, occasionally alternate, sub-coriaceous, linear to elliptic, 2.5 – 3 × 0.5 – 1 cm, apex acute, base rounded, entire, upper and lower surfaces glabrous. Inflorescence 5 – 8 cm long × 10 – 20 mm wide, bracts herbaceous, narrowly triangular, 5 mm long. Calyx 16 mm long, 4-toothed, outer 2 slightly longer than inner 2, with sinus, glabrous but glandular on outer surface (calyx with large extra-floral nectaries scattered), inner surface hairy. Corolla white, hypocrateriform, hairy at throat, tube straight, c. 30 mm long, lobes c. 5 mm across. Stamens lying just below throat, attached towards top of tube. Ovary narrowly flask-shaped, c. 2.5 mm long. Fruit c. 5 mm long, dark brown with reticulate surface, short stylopodium, and without attachment scar. Fig. 24, Plate 3C.

DISTRIBUTION. Minas Gerais (Map 19)
SPECIMENS EXAMINED. MINAS GERAIS: Mun. de Augusto de Lima, Serra do Cabral, 20 March 1994, *Roque et al.* CFCR 15302 (K); Mun. Joaquim Felício, Serra do Cabral, 18 Nov. 1997, *Hatschbach et al.* 67231 (K); Rio Embalassaia, 22 Oct. 1999, *Hatschbach et al.* 69535 (K).
HABITAT. In *cerrado* or *campo rupestre* in sandy soil at c. 1000 m.
CONSERVATION STATUS. Vulnerable. Only four collections, all from the Serra do Cabral which is an unprotected area and likely to suffer a decline in habitat quality in the future due to local extractive activities (Taylor & Zappi 2004).

This taxon has very long corolla tubes.

Group 8 Commutata
Eight species in Brazil.

Key to species in the Commutata group

1. Leaves completely glabrous on upper and lower surface ···················· 63. **S. discolor**
   Leaves hairy on upper and/or lower surface ·· 2
2. Corolla lobes 5 mm across ·· 3
   Corolla lobes 2 mm across ·· 7
3. Inflorescence to 3 cm long ··· 4
   Inflorescence to 7 cm long or more ·· 5
4. Leaves coriaceous, broadly ovate to rotund, base attenuate, upper surface almost glabrous, lower surface with white hairs along nerves, secondary veins forming an intricate reticulate network ··· 62. **S. lacunosa**
   Leaves thickly chartaceous, spathulate, base attenuate, upper and lower surface densely covered with uniseriate hairs, secondary veins barely discernible ························· 57. **S. commutata**
5. Upper leaf surface almost glabrous; inflorescence with rachis visible between flowers; corolla tube 18 – 20 mm long ·· 64. **S. cearensis**
   Upper leaf surface with light or dense covering of long uniseriate hairs; inflorescence dense, rachis not visible between flowers; corolla tube 12 – 16 mm long ································· 6
6. Upper stem, calyx and bracts with only a light covering of short hairs; leaves sub-coriaceous, obovate, 1.5 – 2 × 1 – 1.5 cm, upper surface sparsely covered in long white uniseriate hairs, nerves prominent and dark on undersurface ··· 61. **S. guedesii**

Fig. 24. *Stachytarpheta monachinoi*. **A** habit; **B** flower; **C** corolla; **D** calyx adaxial view; **E** calyx abaxial view; **F** bract. All from *Hatschbach et al.* 67231. DRAWN BY MARGARET TEBBS.

Upper stem, calyx and bracts fairly densely covered with long, patent hairs; leaves chartaceous, obovate, 2.5 – 5.5 × 1 – 3.5 cm, upper and lower surface more or less densely covered with uniseriate hairs; nerves prominent and pale on undersurface ·················· 60. **S. hispida**
7. Upper stem and undersides of leaves so densely hairy as to obscure the surface; hairs yellowish-brown; leaves obovate, 1 – 2 × 0.7 – 1.3 cm; bract shorter than calyx at anthesis ············· 58. **S. viscidula**
Upper stem and undersides of leaves visible between hairs; hairs white; leaves ovate, 2 – 4 × 1 – 2.5 cm; bract as long as or longer than calyx at anthesis ·················· 59 **S. mexiae**

57. S. commutata *Schauer* (1847: 570; 1851: 216). Types: Minas Gerais, 'In umbrosis saxosis montis Itacolumi et alibi in editis montosis prov. Minarum', *Lund* s.n. (isosyntype G), *Sellow* s.n. (isosyntype K).
S. villosa Cham. (1832: 247). Type: Brazil, *Sellow* s.n. (holotype B†; isotype G; photo NY).

Shrub to 50 cm, branched. Stems rounded, densely covered with uniseriate hairs, more densely so towards top of plant. Leaves petiolate, patent or erect, crowded, thickly chartaceous, spathulate, 1 – 3 × 0.8 – 2 cm, apex obtuse to rounded, base attenuate, decurrent into petiole, margin inrolled, crenate in upper part, upper and lower surface densely covered with uniseriate hairs. Inflorescence 2 – 3 cm long × 8 – 10 mm wide; bracts narrowly triangular, c. 10 mm long. Calyx straight, densely covered with uniseriate hairs on outer surface, c. 11 mm long, with 4 subequal teeth and sinus. Corolla blue, hypocrateriform, tube c. 16 mm long, lobes c. 5 mm across. Stamens inserted above middle of tube, ± sessile. Ovary narrow, c. 2 mm long, style c. 21 mm long. Fruit dark brown c. 3 mm long with short stylopodium. Fig. 25A – J.

DISTRIBUTION. Minas Gerais (Map 20).
SPECIMENS EXAMINED. MINAS GERAIS: 1840, *Claussen* s.n. (K); ?date, *Sellow* B 1455 (K); 1874, *Herb. Ferreira* 643 (K); 1883 – 1884, Pico d'Itacolumy, *Glaziou* 15329 (K); Itacolumy, 10 Aug. 1895, *Schwacke* 11530 (G); Itacambira, 29 Nov. 1984, *Kawasaki et al.* SPF 36194/CFCR 6594 (K).
HABITAT. *Campo rupestre* at c. 1200 m.
CONSERVATION STATUS. Endangered. The only modern collection is from Itacambira. There is still some *campo rupestre* left at this locality, but *Eucalyptus* is being planted along the road, displacing parts of the *campo rupestre*. This taxon could be vulnerable.

See notes under *Stachytarpheta villosa* about the misidentification of the *Sellow* specimen by Chamisso.

58. Stachytarpheta viscidula *Schauer* (1847: 570; 1851: 215). Type: Minas Gerais, 'Cachoeira do Campo,' 1840, *Claussen* s.n. (syntypes BR, G) and *Martius* Herbarium number 1044 (isosyntype K).
Shrub to ?75 cm, branched. Stems woody, rounded,

densely covered with uniseriate, patent brownish hairs, often with a globule of ?exudate on top of each hair, becoming more hairy towards top of plant. Leaves sessile, patent, chartaceous, obovate, 1 – 2 × 0.7 – 1.3 cm, apex obtuse to rounded, base cuneate, margin crenate, upper and lower surface densely covered with yellowish-brown uniseriate hairs, more dense on lower surface. Inflorescence c. up to 7 cm long × 15 mm wide, bracts narrowly triangular, 10 mm long. Calyx with 4 ± equal teeth, and deep sinus. Corolla light blue, hypocrateriform, tube straight, c. 15 mm long, inner and outer surface glabrous, lobes c. 2 mm across. Stamens lying at middle of tube. Ovary narrow, c. 2 mm long, style exserted. Fruit dark brown, c. 5 mm long without prominent attachment scar.

DISTRIBUTION. Minas Gerais. Known only from the types.
HABITAT. 'In mountainous places'

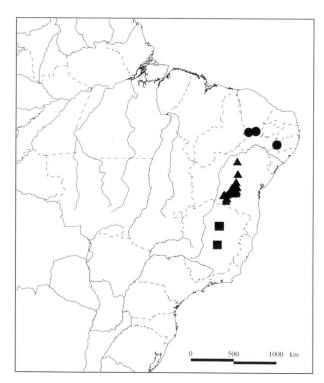

Map 20. Distributions of *Stachytarpheta cearensis* (●), *S. commutata* (■) and *S. hispida* (▲).

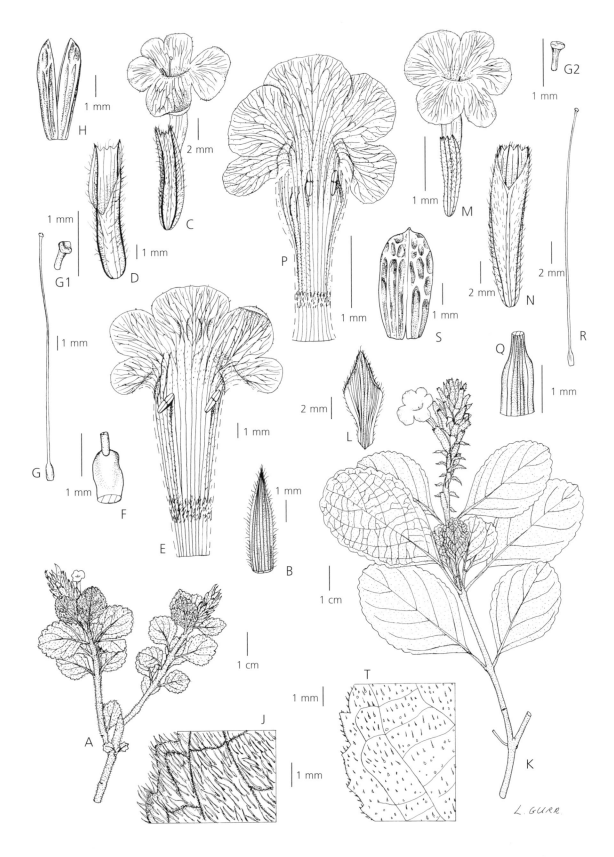

Fig. 25. *Stachytarpheta commutata.* **A** habit; **B** bract; **C** calyx and corolla; **D** calyx; **E** corolla; **F** ovary; **G** style and stigma; **H** fruit; **J** detail of upper leaf surface. *Stachytarpheta cearensis.* **K** habit; **L** bract; **M** calyx and corolla; **N** calyx; **P** corolla; **Q** ovary; **R** style and stigma; **S** fruit; **T** detail of upper leaf surface. **A – J** from *Kawasaki et al.* SPF 36194/CFCR 6594; **K – T** from *Andrade et al.* 238. DRAWN BY LINDA GURR.

CONSERVATION STATUS. Critically endangered. It is extremely difficult to pinpoint the localities for this taxon. I have found no modern collections.

This is a much more viscid plant than *Stachytarpheta commutata*, and there is evidence on the type specimen that the inflorescence can reach 7 cm long, as opposed to only 3 cm long in *S. commutata*. The leaf shape is distinctly ovate and not spathulate.

59. S. mexiae *Moldenke* (1940: 472). Type: Minas Gerais, Diamantina, *Mexia 5824* (holotype NY n.v.; isotype K).

Shrub to 3.3 m, branched. Stems rounded, fairly densely covered with white uniseriate hairs. Leaves petiolate, patent, stems with long internodes, chartaceous, ovate, 2 – 4 × 1 – 2.5 cm, apex acute to obtuse, base truncate, then tapering into petiole, upper surface covered with uniseriate hairs, bullate, lower surface more densely hairy. Inflorescence to 14 cm long × 10 mm wide; bracts narrowly triangular, densely covered with uniseriate hairs, c. 10 mm long. Calyx straight, hairy, c. 10 mm long, irregularly 4-toothed with sinus. Corolla blue, hypocrateriform, tube c. 10 mm long, lobes c. 2 mm across. Stamens inserted below middle of tube, anthers lying just above middle. Ovary narrow, c. 2 mm long; style c. 20 mm long. Fruit dark brown to black, c. 6 mm long.

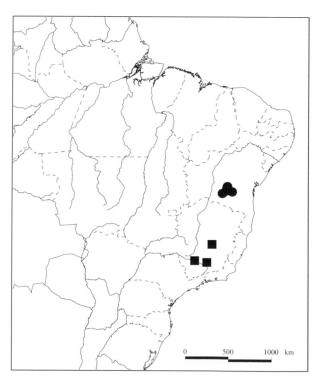

Map 21. Distributions of *Stachytarpheta lacunosa* (●) and *S. mexiae* (■).

DISTRIBUTION. Minas Gerais (Map 21).
SPECIMENS EXAMINED. MINAS GERAIS: Diamantina Distr., July 1840, *Gardner 5113* (K); Serra do Cipó, 16 Feb. 1968, *Irwin et al. 20224* (NY); 27 Oct. 1988, *Harley et al. 25421* (K); Mun. Santa Bárbara, Serra do Brucutú, 6 Feb. 1943, *Magalhães 2736* (NY); Serra de Ouro Preto, 13 Sept. 1893, *Schwacke 9406* (G); Mun. Tiradentes, Serra de São José, 16 Jan. 1994, *Atkins et al. CFCR 13667*; Mun. Serro, Boca da Mata, 6 Oct. 1945, *Williams & Assis 7920* (NY).
HABITAT. Open slopes at 1100 – 1200 m.
CONSERVATION STATUS. Near threatened. This taxon is to be found at middle altitudes along the Serra do Espinhaço, Minas Gerais, from Diamantina to Tiradentes. It is not thought to be vulnerable. The localities within the Parque Nacional da Serra do Cipo are protected.

This is different from previous species in this section, as it can become a tall shrub. The type specimen says to 3.3 m, and where given, the height of the other specimens ranges from 1 – 2 m. It is not viscid, has ovate leaves and inflorescences to 14 cm long.

60. S. hispida *Nees & Mart.* (1823: 69); Schauer (1847: 569; 1851: 214); Atkins in Stannard (1995). Type: Bahia, Barra da Vareda in Campis, *Prince Maximilian of Wied* s.n. (isotype BR).

Small shrub to 80 cm, branched. Stems rounded, ± densely covered with uniseriate hairs, more dense on younger growth, hairs normally white, occasionally brownish. Leaves petiolate, patent to erect, chartaceous, obovate, 2.5 – 5.5 × 1 – 3.5 cm, apex obtuse to rounded, base attenuate, decurrent into petiole, margin crenate above middle, upper and lower surface ± densely covered with uniseriate hairs. Inflorescence to 7 cm × c. 10 mm; bracts linear, c. 12 mm long. Calyx straight, c. 15 mm long, with 4 ± equal teeth, and deep sinus, sparsely covered on outer surface with uniseriate hairs. Corolla light or lavender blue, hypocrateriform, tube slightly kinked, c. 16 mm long, lobes c. 5 mm across. Stamens inserted at middle of tube, anthers lying in upper part. Ovary oblong, c. 2 mm long; style c. 20 mm long. Fruit pale brown c. 6 mm long. Plate 3B.

DISTRIBUTION. Bahia (Map 20).
SPECIMENS EXAMINED. BAHIA: Serra do Curral Feio, 6 March 1974, *Harley et al. 16849* (K); Serra Geral de Caitité, 9 April 1980, *Harley et al. 21124* (K); 12 April 1980, *Harley et el. 21264* (K); Mun. de Caetité, 18 Feb. 1992, *de Carvalho et al. 3696* (K); 12 March 1994, *Roque et al. CFCR15004*; Serra do Rio de Contas, 16 Jan. 1974, *Harley et al. 15136* (K); 24 March 1977, *Harley et al. 19940* (K); Mun. Rio de Contas, Pico das

Almas, 18 March 1977, *Harley et al.* 19655 (K); 21 Feb. 1987, *Harley et al.* 24530 (K); 28 Oct. 1988, *Harley et al.* 25730; 30 Nov. 1988, *Harley et al.* 26505 (K); 3 March 1994, *Atkins et al.* CFCR 14794; 16 Nov. 1996, *Bautista et al.* PCD 4358; Rio de Contas, 8 Nov. 1988, *Harley et al.* 26012 (K); 4 March 1994, *Sano et al.* CFCR 14857 (K); Mun. Piatã, 21 Dec. 1984, *Furlan et al.* SPF 37190/CFCR 7390 (K); 15 Feb. 1987, *Harley et al.* 24264 (K); 7 Jan. 1992, *Harley et al.* H50681; 23 Feb. 1994, *Atkins et al.* CFCR 14464; Mun. Morro do Chapéu, 1 June 1980, *Harley et al.* 22942 (K); Barra da Estiva, 4 July 1983, *Coradin et al.* 6433 (K); Mun. Abaíra, 15 Feb. 1992, *Harley et al.* 52088; 7 Feb. 1992, *Stannard et al.* H51062; 13 March 1992, *Stannard et al.* H51901; Mun. Palmeiras, 19 Feb. 1994, *Atkins et al.* CFCR 14234.

HABITAT. *Campo rupestre* and *campo geral*, at 1000 – 1600 m.

CONSERVATION STATUS. Least concern. A common species at upper altitudes of the Chapada Diamantina from Caetité in the south to Morro do Chapeu in the north. The creation of the Parque Nacional da Chapada Diamantina makes this a protected plant within that area.

The isotype from BR has a Herbarium Martius label with 'ad Tamburil et Valos in campis' written on it. It has another, rougher label in the right hand corner with 'B.d.V.' written on it. I take this to mean Barra de Vareda, and is probably the original collecting label.

The species varies in leaf shape and size, and density of indumentum. *Ganev* 927 may fit here. It was collected from Mun. Piatã, Serra do Atalho, near Garimpo da Cravada. It has thicker leaves, more like *Stachytarpheta crassifolia*, but in all other respects is like *S. hispida*. *Melo* 981, collected from Serra do Atalho, has much longer inflorescences, larger leaves, and is much hairier, but also probably belongs here.

This species differs from the previous ones in this section by its large corolla lobes, and by its distribution.

61. Stachytarpheta guedesii *S. Atkins* **sp. nov.** a *S. lacunosa* bracteis cum calycibus et caulibus superne foliisque pilis longis albis subtiliter obsitis (non glabris), inflorescentia usque 7 cm longa (non 1.5 – 2 cm) differt. Typus: Bahia, Mun. Palmeiras, Pai Inácio, caminho para o Cercado, 29 June 1995, *Guedes et al.* PCD 2010 (holotypus ALCB; isotypus K).

Subshrub to 40 cm, branched. Stem 4-sided, covered in short, white hairs, more numerous towards top of plant, and at nodes. Leaves subcoriaceous, obovate, 1.5 – 2 × 1 – 1.5 cm, apex rounded, base attenuate, decurrent into petiole, upper surface covered in long white uniseriate hairs, nerves slightly impressed,

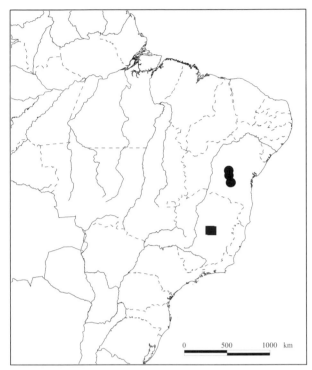

Map 22. Distributions of *Stachytarpheta discolor* (■) and *S. guedesii* (●).

lower surface with reticulate network of nerves, whole surface densely covered with uniseriate hairs. Inflorescence up to 7 cm long × 10 mm wide, previous year's inflorescence normally present; bracts narrow, 8 mm long, hairy, woody. Calyx straight, c. 10 mm long, slightly hairy on outer surface, 4-toothed with sinus on adaxial side. Corolla mauve, hypocrateriform, tube straight, c. 12 mm long, lobe c. 5 mm across, fine hairs at throat. Stamens attached at middle, ± sessile. Ovary flask-shaped, c. 2 mm long, style c. 12 mm long. Fruit dark brown, c. 6 mm long without stylopodium. Fig. 26G – M.

DISTRIBUTION. Bahia (Map 22).

SPECIMENS EXAMINED. BAHIA: Barra da Estiva, Morro do Ouro, 19 Nov. 1988, *Harley et al.* 26932 (K); 23 Nov. 1992, *Arbo et al.* 5720 (K); 16 Feb. 1997, *Passos et al.* PCD 5775 (K); Mun. Lençóis, estrada para Cercado, 23 Aug. 1996, *Harley & Mayworm* PCD 3763 (K); Mun. Mucugê, caminho para Guiné, 15 Feb. 1997, *Guedes et al.* PCD 5666 (K); Mun. Palmeiras, Pai Inácio, 28 Dec. 1994, *Guedes et al.* PCD 1423 (K).

HABITAT. *Campo rupestre* and *campo geral* at c. 1000 m.

CONSERVATION STATUS. Near threatened. This is not a common species. It is found along the Chapada Diamantina from Mucugê in the south to Palmeiras in the north. Some localities may be within the protected area of the Parque Nacional de Chapada Diamantina, but those areas which fall outside are under pressure from encroaching agriculture.

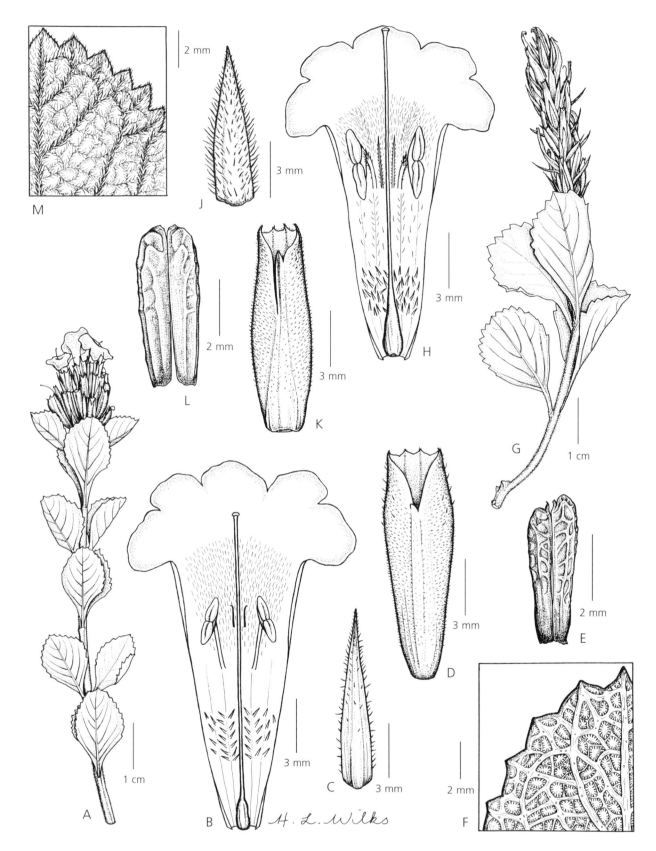

Fig. 26. *Stachytarpheta lacunosa.* **A** habit; **B** corolla; **C** bract; **D** calyx; **E** fruit; **F** detail of underside of leaf. *S. guedesii.* **G** habit; **H** corolla; **J** bract; **K** calyx; **L** fruit; **M** detail of underside of leaf. **A – D** & **F** from *Ganev* 500; **E** from *Atkins et al.* CFCR 14687; **G** & **M** from *Guedes et al.* 2010; **H – J** from *Arbo et al.* 5720. DRAWN BY HAZEL WILKS.

This is superficially very similar to *Stachytarpheta lacunosa*, but the bracts, calyces and upper leaves are more hairy, and the inflorescence is much longer.

This species is named for Lenise Guedes who collected the type specimen.

62. Stachytarpheta lacunosa *Mart. ex Schauer* (1847: 571; 1851: 218). Type: Bahia, 'in campis editis prov. Bahiensis interioris in Brasilia', *Martius* s.n. (holotype M n.v.; photo from Munich Herbarium in NY).

S. reticulata var. *bahiensis* Moldenke (1983b: 399). Type: Brazil, Bahia, Mun. Rio de Contas, 17 May 1983, *Hatschbach* 46523 (holotype TEX; isotype MBM).

Subshrub to 60 – 70 cm (– 1.2 m), branched. Stems 4-sided, woody and gnarled, glabrous towards base of plant, becoming hairy with short uniseriate hairs towards apex. Leaves sessile, erect, coriaceous, broadly ovate to rotund, 1 – 2 × 0.8 – 1.8 cm, apex rounded, base attenuate, margin serrate, upper surface almost glabrous, lower surface with white hairs along nerves, secondary veins forming an intricate reticulate network. Inflorescence 1 – 2.5 cm × 10 – 20 mm wide; bracts narrowly subulate, 6 – 7 mm long, almost glabrous except along margin. Calyx 10 mm long, with 4 adaxial teeth, sparsely hairy on outer surface with uniseriate hairs; corolla blue with white throat, hypocrateriform, c. 10 mm long, almost glabrous at throat, lobes c. 5 mm across. Stamens attached at middle, anthers lying just above middle. Ovary oblong, c. 1 mm long, style 15 mm long. Fruit dark brown, reticulate, 5 mm long, without stylopodium. Fig. 26A – F, Plate 8B.

DISTRIBUTION. Bahia (Map 21).
SPECIMENS EXAMINED. BAHIA: Mun. Piatã, Povoado da Tromba, 1000 m, 15 June 1992, *Ganev* 500 (HUEFS; K); Mun. de Abaíra, distrito de Catolês, Boa Vista, 1250 m, 5 May 1992, *Ganev* 239 (K); Boa Vista, acima do Capão do mel, 1250 m, 11 June 1994, *Ganev* 3342 (K); Mun. Rio de Contas, estrada para Arapiranga, 950 – 1050 m, 11 Nov. 1988, *Harley et al.* 26114 (K); Road between Mato Grosso and the foot of Morro do Itabira, 1 March 1994, *Atkins et al.* CFCR 14687.
HABITAT. *Campo rupestre*, on rocky, quartzitic or white sandy soil between 1000 and 1600 m.
CONSERVATION STATUS. Vulnerable. All localities are within a 25 km radius of each other, along the summits of the northern end of the Chapada Diamantina. At present these areas are not protected.

Very similar to the previous species. See note there about differences. The present species grows at higher altitudes, and occurs in a limited area.

63. Stachytarpheta discolor *Cham.* (1832: 251). Schauer (1847: 571; 1851: 218). Type: without locality, *Sellow* (holotype B†; photo from Berlin Herb. NY).

Shrub to 1 m, branched. Stems rounded, glabrous, with short tufts of white hairs around nodes. Leaves sessile, patent, ovate, 2.2 – 2.8 × 1.3 – 2 cm, apex obtuse, base truncate, upper and lower surface glabrous, but uniformly covered with sessile glands, interspersed with large dark nectaries, margin coarsely serrate. Inflorescence 1.4 – 4 cm long × 12 – 15 mm wide (note: 2 lateral branches arise each side of the inflorescence, so inflorescences are tucked between 2 longer branches, giving them the appearance of being lateral). Bracts narrowly triangular, c. 7 mm long. Calyx 8 mm long, 4-toothed with sinus on adaxial side, outer surface glabrous with sessile glands and one or two dark nectaries. Corolla dark blue, infundibular, tube straight, c. 12 mm long, hairy at throat, lobes c. 2 mm across. Stamens inserted just below throat. Ovary narrowed at top, c. 1 mm long; style 15 mm long. Fruit dark brown, reticulate, 4 mm long. Fig. 27.

DISTRIBUTION. Minas Gerais (Map 22).
SPECIMENS EXAMINED. MINAS GERAIS: Mun. Tijucal, Trinta Reis, 4 km N, (on Serro-Datas Road, pers. comm. F. Salimena Pires), 13 March 1982, *Hatschbach* 44687 (NY); Mun. Serro, Cascata Moinho de Esteira, 25 Oct. 1999, *Hatschbach et al.* 69712 (K); Mun. Presidente Kubitschek, estrada Serro-Datas, 18 Feb. 2003, *França et al.* 4565 (HUEFS, K).
HABITAT. In *campo rupestre*, among rocks.
CONSERVATION STATUS. Vulnerable. Not a widespread or common species. Outcrops of campo rupestre in this area are isolated. Often the areas are grazed by cattle. This species could be vulnerable to further encroachment.

This species is different from most of the others in this section because the stem and leaves are glabrous.

64. Stachytarpheta cearensis *Moldenke* (1941b: 54). Type: Ceará, without date, *Allemão* 1152 (holotype R n.v.; photo of type NY).

Shrub to 2.5 m. Stem 4-sided, covered in short white hairs. Leaves petiolate, chartaceous, ovate, 2.5 – 7 × 1.5 – 4.5 cm, apex obtuse, base attenuate, decurrent into petiole, crenate, upper surface glabrous to sparsely strigose, lower surface tomentose. Inflorescence 5 – 9 cm long by 15 – 20 mm wide; bracts subulate, c. 9 mm long, sparsely covered with uniseriate hairs. Calyx 14 mm long, 4-toothed, teeth equal with deep sinus, sparsely covered with uniseriate hairs. Corolla bluish mauve with white throat, tube 18 – 20 mm long, lobes c. 5 mm across. Stamens lying just below throat, attached towards top of tube. Ovary flask-shaped, c. 2 mm long. Fruit dark brown, without prominent attachment scar, c. 3 mm long. Fig. 25K – T.

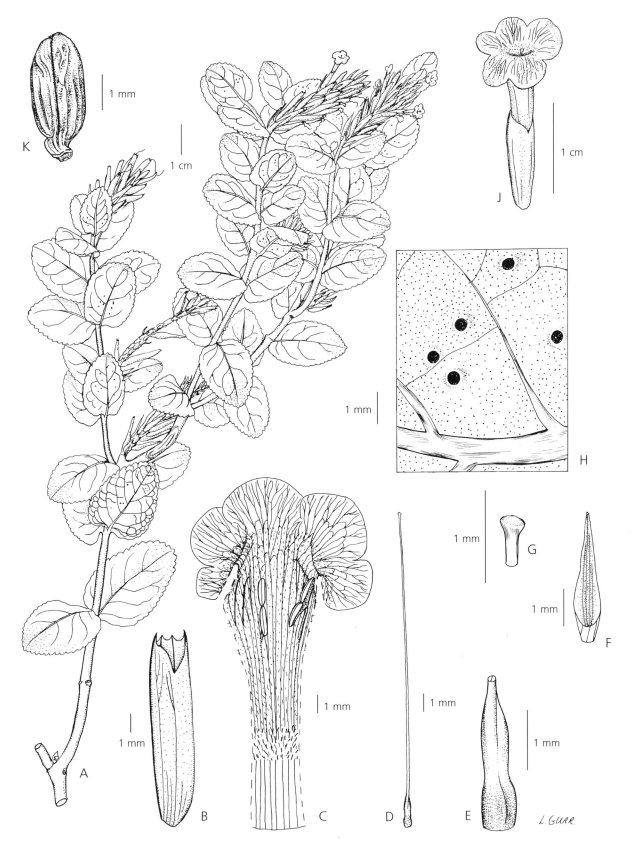

Fig. 27. *Stachytarpheta discolor.* **A** habit; **B** calyx; **C** corolla; **D** gynoecium; **E** ovary; **F** bract; **G** stigma; **H** detail of underside of leaf; **J** flower; **K** fruit. From *França et al.* 4557. DRAWN BY LINDA GURR.

DISTRIBUTION. Ceará and Pernambuco (Map 20).
SPECIMENS EXAMINED. CEARÁ: Mun. de Crato, Floresta Nacional do Araripe, 6 May 1991, *Esteves & Barros* 2581 (K). **PERNAMBUCO:** Buíque, Serra do Catimbau, 18 Aug. 1994, *Rodal* 262 (PEVFR, K); 18 Oct. 1994, *Sales* 429; 17 Aug. 1995, *Figueiredo et al.* 799; 10 Oct. 1995, *Andrade et al.* 238 (PEVFR, K).
HABITAT. Humid forest and dense shrubbery.
CONSERVATION STATUS. Near threatened. The National Forest of Araripe in Ceará is a protected area. The

'brejos de altitude de Pernambuco' in the area around Buique has been studied by Sales *et al.* (1999) and found to have a distinctive vegetation, as yet unprotected.

Ule 7194 (K)- Piauí, Serra Branca is possibly a form of *Stachytarpheta cearensis.*

Group 9 Sellowiana
Three species in Brazil.

Key to species in the Sellowiana group

1. Bracts to 2 mm wide at widest point just above base, tapering above; corolla tube to 20 mm long · 66. **S. ajugifolia**

   Bracts no wider than 1 mm, sides ± parallel to apex; corolla tube to 15 mm long · · · · · · · · · · · · · · · · · · · 2
2. Inflorescence to 3 cm wide; calyx and bracts covered with gland-topped hairs; calyx to 14 mm long · 65. **S. sellowiana**

   Inflorescence to 1 cm wide; calyx and bracts without gland-topped hairs; calyx to 10 mm long · 67. **S. itambensis**

65. Stachytarpheta sellowiana *Schauer* (1847: 571; 1851: 218); Atkins *et al.*(1996: 33 – 35). Type: In stony places, Serra de São José, Prov. S. Pauli (sic), alt. 4000' [1200 m], June, *Riedel* s.n. (syntype B†, LE n.v.); location unknown *Sellow* s.n., (probable isosyntype K).

Subshrub to 1 m, branched. Stems rounded, woody, gnarled, covered in minute gland-topped hairs. Leaves sessile, coriaceous, patent, obovate, 2 – 5 × 1 – 2 cm, apex obtuse, base attenuate, serrate, upper and lower surface fairly densely covered with very short simple hairs, interspersed with minute gland-topped hairs. Inflorescence 3 – 6 × 3 cm; bracts narrow-ovate, c. 8 mm long, finely covered with short hairs, some gland-topped. Calyx 2-lipped, c. 14 mm long covered with gland-topped hairs; corolla infundibular, dark blue, tube slightly curved, c. 15 mm long, lobes c. 3 mm across. Stamens lying half way down tube. Ovary flask-shaped, c. 2 mm long. Fruit not seen. Fig. 28A – D, Plate 4B.

DISTRIBUTION. Minas Gerais.
SPECIMENS EXAMINED. MINAS GERAIS: Without locality, *Sellow* s.n.; Serra de São José, Tiradentes, Dec. 1893, *Schwacke* 10104 (G); *Schwacke* 10105 (G); 25 April 1992, *Alves* 4005 (K); 16 Jan. 1994, *Atkins et al.* CFCR 13680 (K).
HABITAT. In stony places at 1120 – 1200 m.
CONSERVATION STATUS. Critically endangered. This species is known from a single serra near Tiradentes. When I visited, only six plants were found. (Atkins *et al.* 1996). This species is Critically Endangered.

The glandular hairs on the stem, leaves, calyx and corolla distinguish this species different from all others in this section.

66. Stachytarpheta ajugifolia *Schauer* (1847: 570; 1851: 215). Type: Minas Gerais, in stony places in the Serra de São José, prov. S. Pauli (sic), 4000' [1200 m], *Riedel* s.n. (syntype LE n.v.); *Ackermann* s.n. (isosyntype BR).

Stout shrub to 1 m, branched. Stems sub-terete, densely covered with uniseriate hairs. Leaves ± sessile, erect, chartaceous, oblong to elliptic, 3 – 4 × 1.4 – 2.5 cm, apex obtuse, base truncate to short attenuate, crenate, upper surface strigose, lower surface tomentose. Inflorescence 2 – 5 cm long × 10 mm wide, bracts sub-woody, linear, c. 10 mm long, densely covered with uniseriate hairs. Calyx 15 mm long, densely covered with uniseriate hairs, 4-toothed, teeth ± equal, outer pairs twisting together, with deep sinus between, teeth long and narrow. Corolla dark blue, hypocrateriform, not hairy at throat, tube straight, 20 mm long, lobes c. 3 mm across. Stamens attached at middle of tube, anthers lying just above middle. Ovary narrow, c. 2 mm long. Fruit dark brown, with short stylopodium.

DISTRIBUTION. Minas Gerais (Map 23).
SPECIMENS EXAMINED. MINAS GERAIS: Without locality, *Claussen* 205 (G, BR); Serrinha prope Santa Rica de Ibitipoca, 12 Aug. 1896, *Schwacke* 12346 (G); Serra de Ouro Preto, 29 Aug. 1898, *Schwacke* 13465 (G); 13

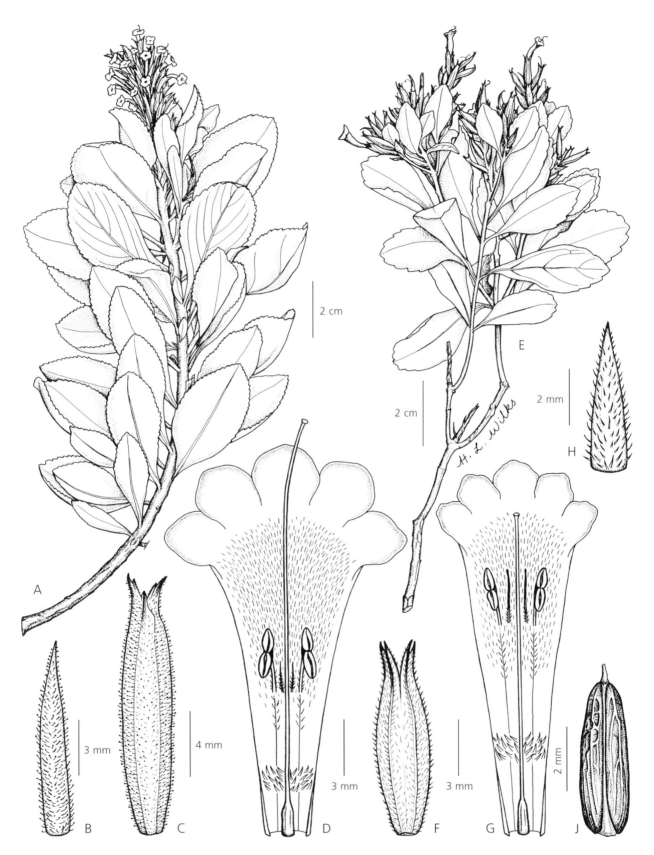

Fig. 28. *Stachytarpheta sellowiana.* **A** habit; **B** bract; **C** calyx; **D** corolla. *Stachytarpheta itambensis.* **E** habit; **F** calyx; **G** corolla; **H** bract; **J** fruit. **A – C** from *Atkins et al.* CFCR 13680; **D** from *Alves* 4005; **E** and **J** from *Anderson et al.* 35918; **F – H** from *Furlan et al.* SPF 23281/CFCR 3074. DRAWN BY HAZEL WILKS.

Sept. 1893, *Schwacke* 9406 (G); Tiradentes, Serra de São José, 24 April 1992, *Alves et al.* 4025 (K).

HABITAT. No information, only one altitude, 1015 m.

CONSERVATION STATUS. Data deficient. Only 3 collections, and no habitat information. These localities are at the southernmost tip of the Serra do Espinhaço range, close to the state border with Rio de Janeiro. There are no specifically protected areas in this area, and this taxon is vulnerable.

This species is similar to *Stachytarpheta mexiae* from the previous group, but differs in its shorter inflorescence, and its calyx with a sinus on the abaxial side.

67. Stachytarpheta itambensis *S. Atkins* **sp. nov.** a *S. sellowiana* foliis plerumque minoribus 2 – 3.5 × 1 – 1.5 (non 2 – 5 × 1 – 2 cm) infra reticulariter venosis pilis brevibus uniseriatis dense obsitis (non pilis brevissimis simplicibus moderate obsitis cum pilis glandulis terminatis interspersis), calyce modo pilis brevibus uniseriatis obsito (sine pilis glandulis terminatis interspersis) distincta. Typus: Minas Gerais, Morro do Pico Itambé, 6 April 1982, *Furlan et al.* SPF 23281/CFCR 3074 (holotypus SPF; isotypus K).

S. *viscidula* var. *brevipilosa* Moldenke (1973a: 222). Type: Minas Gerais, eastern slopes of Pico do Itambé, 13 Feb. 1972, *Anderson et al.* 35918 (holotype TEX; isotype K).

S. *lacunosa* var. *angustifolia* Moldenke (1973a: 221). Type: Minas Gerais, Pico Itambé, 9 Aug. 1972, *Hatschbach* 30115 (holotype TEX).

Shrub to 1.5 m. Stems rounded, upper branches 4-sided, fairly densely covered in very short, uniseriate hairs. Leaves crowded, short-petiolate, sub-fleshy, ovate, 2 – 3.5 × 1 – 1.5 cm, apex obtuse, base attenuate, margin crenate, upper and lower surface reticulately veined, lower surface densely covered with short, uniseriate hairs, upper surface less dense, hairs of upper surface with bulbous bases. Inflorescence c. 2 × 1 cm; bracts narrowly triangular, c. 4 mm long, quite hairy. Calyx c. 10 mm long, hairy, with 4 equal teeth and sinus. Corolla blue, hypocrateriform, tube straight, c. 15 mm long. Stamens in upper part of tube. Ovary c. 1 mm long, style, c. 2.4 mm long, exserted. Fruit dark brown, reticulate, without stylopodium and attachment scar. Fig. 28E – J.

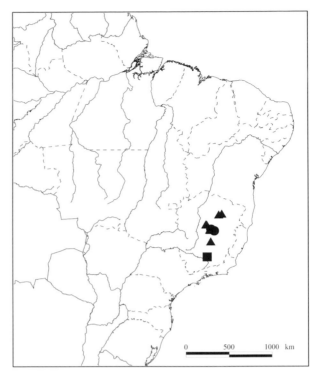

Map 23. Distributions of *Stachytarpheta ajugifolia* (■), *S. harleyi* (▲) and *S. itambensis* (●).

DISTRIBUTION. Minas Gerais (Map 23).

SPECIMENS EXAMINED. MINAS GERAIS: Mun. Diamantina, Rodoviario Guinda-Conselheiro Mata, 16 March 1987, *Hatschbach et al.* 50954 (K).

HABITAT. On steep slopes, with woods and rocky outcrops at 1300 – 3000 m.

CONSERVATION STATUS. Endangered. Four collections from a small area. The area around Diamantina is not protected although an area of very high endemism (Giulietti & Pirani 1997).

A smallish shrub to 50 cm, similar to *Stachytarpheta sellowiana*, but with smaller leaves, and without gland-tipped hairs. *França et al.* 4505, collected from the same area as *Hatschbach et al.* 50954 probably fits here, but the undersides of the leaves are less reticulate and less hairy.

Group 10 Glabra
Five species in Brazil.

Key to species in the Glabra group

1. Leaves smaller than 6 – 14 × 2 – 6 mm, lamina with indumentum of stiff, white hairs; inflorescence
   3 – 6-flowered; bracts almost as long as calyx · 72. **S. diamantinensis**
   Leaves larger than 6 –14 × 2 – 6 mm, lamina ± glabrous; inflorescence 8 – 22-flowered; bracts shorter
   than calyx · 2

2. Leaves broadly ovate, (2 –)3.5 – 4.5 × (1.5 –)3 – 3.8 cm, apex obtuse to rounded, base attenuate · · · · · · · · ·
    · **71. S. cassiae**
    Leaves oblong, elliptic to broadly elliptic, 1.5 – 5 × (0.8 –) 1 – 1.4 cm, apex acute, base long-attenuate · · · · 3
3. Rachis, bracts and calyx covered with short, white hairs · **70. S. pohliana**
    Rachis, bracts and calyx glabrous · 4
4. Leaves oblong, 1.5 – 3 × 0.5 – 1.2 cm; bracts ovate, c. 6 mm long · · · · · · · · · · · · · · · · · **69. S. harleyi**
    Leaves elliptic to broadly elliptic, 3 – 6 (10) × 1.2 – 2 (5) cm; bracts linear to narrowly triangular,
    7 – 10 mm long · **68. S. glabra**

---

68. Stachytarpheta glabra *Cham.* (1832: 250); Schauer (1847: 568; 1851: 211). Type: 'e Brasilia', *Sellow* s.n. (holotype B†; probable isotype K).

S. glabra var. *latifolia* Schauer (1847: 568). Type as above. See below.

S. glabra var. *angustifolia* Schauer (1847: 568). Type: In Prov. Minarum, *Martius* s.n. (n.v.). See below.

Branched shrub to 1.5 m. Stems woody, rounded, glabrous. Leaves petiolate, patent, chartaceous, elliptic to broadly elliptic, 3 – 6 (– 10) × 1.2 – 2 (– 5) cm, apex acute, base long-attenuate into petiole, margin crenate-serrate, upper and lower surfaces glabrous, with occasional scattered white hairs. Inflorescence up to 7 cm long by 12 – 15 mm wide; bracts herbaceous, linear to narrowly triangular, 7 – 10 mm long. Calyx c. 12 mm long, 4-toothed abaxially with outer teeth longer than inner teeth, sinus adaxially, inner and outer surface glabrous. Corolla blue, hypocrateriform, tube straight, 12 – 20 mm long, widening slightly above middle, inner surface hairy at throat with glandular hairs at base; lobes small, regular, c. 2 mm across. Stamens inserted at upper part of tube, anthers at throat. Ovary 3 mm long; style 20 mm long (exserted). Fruit dark brown, surface somewhat reticulate, prominent attachment scar, with stylopodium. Fig. 29F – K.

DISTRIBUTION. Minas Gerais (Map 24).

SPECIMENS EXAMINED. MINAS GERAIS: Itabira do Campo, Feb. 1839, *Claussen* s.n. Herb. Martius 1027 (BR, K); Mun. Diamantina: Serra do Rio Grande, 4 May 1931, *Mexia* 5753 (K); 6 km N de Diamantina, camino a Biribiri, 8 April 1982, *Hensold et al.* SPF 23347/CFCR 3139 (K) & 14 Feb. 1991, *Arbo et al.* 5036 (K); Estrada da Sopa para São José da Chapada, 12 Dec. 1980, *Menezes et al.* SPF 21445/CFCR 545 (K); Mun. de Ouro Preto: c. 4 km NE of Ouro Preto, near Cachoeira das Andorinhas, 29 Nov. 1965, *Eiten* 7001 (K); Proximo a Vila Residencial Samarco, 16 Jan. 1994, *Atkins et al.* CFCR 13800 (K); Mun. Alvorada de Minas, alrededores de Itapanhoacan, 17 May 1990, *Arbo et al.* 4313 (K).

HABITAT. On *campo rupestre*, often on quartzitic outcrops, above 800 m.

CONSERVATION STATUS. Least concern. A fairly common and widespread species.

Quite large variation in leaf size and shape, as noted by Schauer (1851). From Chamisso's (1832) original description, the original specimens seen had elliptic leaves with acutely serrate margins, but he makes no observations about variations. Schauer (1847) made 2 informal groups: α. *latifolia* with elliptic leaves and with coarsely serrate margins, and β. *angustifolia* with lanceolate leaves with remote and shortly serrate margins. He assigns Chamisso's concept to the former. I have interpreted *Stachytarpheta glabra* rather broadly, as the majority of specimens have elliptic to broadly elliptic leaves, and there appears to be no geographic dividing line between the different leaf shapes.

Map 24. Distributions of *Stachytarpheta cassiae* (●) and *S. glabra* (■).

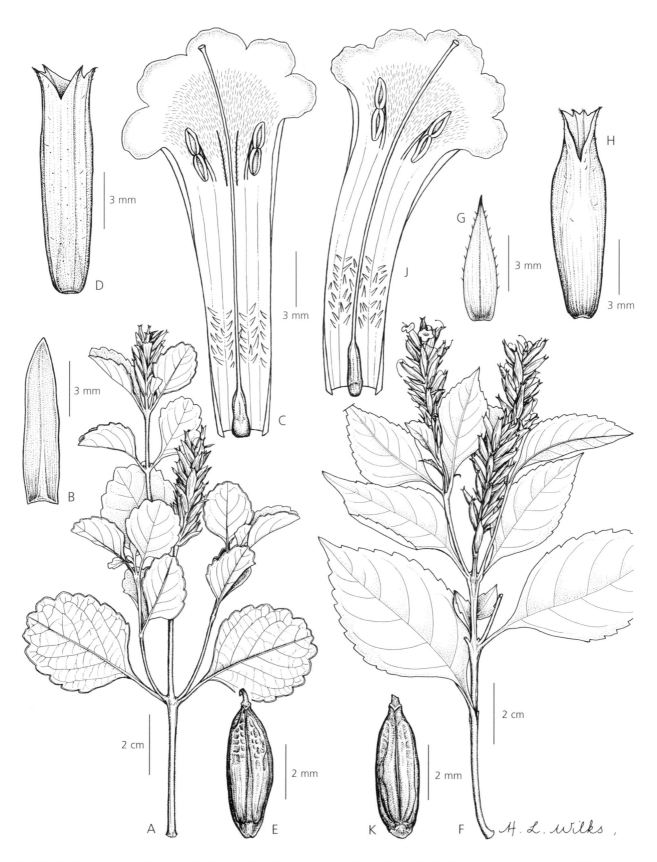

Fig. 29. *Stachytarpheta cassiae.* **A** habit; **B** bract; **C** corolla; **D** calyx; **E** fruit. *S. glabra.* **F** habit; **G** bract; **H** calyx; **J** corolla; **K** fruit. **A – E** from *Sakuragui et al.* CFCR 15086; **F – H** from *Irwin et al.* 29635; **K** from *Hensold et al.* SPF 22305. DRAWN BY HAZEL WILKS.

69. **Stachytarpheta harleyi** *S. Atkins* **sp. nov.** a *S. martiana* foliis oblongis (non obovatis) minoribus 1.5 – 3 × 0.5 – 1.2 cm (non 3.5 – 7 × 1.5 – 2.8 cm), inflorescentiae longitudine 2 – 3 cm (non 16 mm) differt. Typus: Minas Gerais, Mun. Joaquim Felicio, Serra do Cabral, 15 May 2001, *Hatschbach et al.* 72073 (holotypus MBM; isotypus K).

Shrub to 2.5 m. Stems woody, rounded to 4-sided, glabrous. Leaves sessile, subcoriaceous, oblong, 1.5 – 3 × 0.5 – 1.2 cm, apex acute, base attenuate, shallowly crenate, slightly inrolled, upper and lower surface glabrous. Inflorescence 2 – 3 cm long × 10 mm wide; bracts ovate, c. 6 mm long, glabrous. Calyx c. 10 mm long, glabrous, with 4 teeth and sinus. Corolla hypocrateriform, blue, tube 20 mm long, lobes c. 4 mm across. Stamens lying just below throat, attached towards top of tube. Ovary narrow oblong, c. 2 mm long, style 18 mm long. Fruit black, reticulate, with short stylopodium. Fig 30A – F.

DISTRIBUTION. Minas Gerais (Map 23).
SPECIMENS EXAMINED. MINAS GERAIS: Serra do Cabral, 16 Jan. 1996, *Hatschbach & Silva* 64318 (K) & 16 March 1997, *Hatschbach et al.* 66320 (K). Diamantina, 24 km na estrada Diamantina-Conselheiro Mata, 30 Aug. 1981, *Giulietti et al.* SPF 21616/CFCR 1835 (K); 17 Feb. 2003, *França et al.* 4543 (HUEFS; K); Ao norte de Grão Mogol, 27 Nov. 1984, *Harley et al.* SPF 36078/CFCR 6478 (K).
HABITAT. *Campo rupestre* at 900 – 1276 m.
CONSERVATION STATUS. Vulnerable. Collections from the Serra do Cabral which is an area not protected and likely to suffer a decline in habitat quality in the future due to local extractive activities (Taylor & Zappi 2004).

This species is named for R. M. Harley who has been the inspiration for this work.

70. **Stachytarpheta pohliana** *Cham.* (1832: 248); Schauer (1847: 568; 1851: 210). Type e Brasilia, Sellow (holotype B†; possible isotype K — the specimen in type cover at Kew is ex museo botanico Berolinensi and has *Sellow* 1456, but no number is cited by Chamisso. In *Flora Brasiliensis* Schauer says "In monte Itambé prov. Minarum: *Sellow.*")

Shrub to 60 cm. Stems woody, rounded, sparsely covered with white, uniseriate hairs, more numerous towards apex of plant. Leaves petiolate, patent, chartaceous, broadly elliptic to ovate, 2.5 – 6.5 × 1 – 2.5 cm, apex acute, base attenuate, margin crenate, upper and lower surface with uniseriate hairs along main vein, and with scattered uniseriate hairs over surface. Inflorescence 3 – 4 × 1.5 cm; bracts narrowly

triangular, c. 6 mm long, striate, with ciliate margin. Calyx c. 10 mm long, with scattered uniseriate hairs, 5-toothed, with 4 ± equal, and 1 very small on anterior side. Corolla infundibular, deep blue, (no more information from specimens: rest from original description), glabrous on outer surface, subsericeous at throat. Stamens inserted in upper part of tube. Fruit dark brown, reticulate, small stylopodium, without prominent attachment scar.

DISTRIBUTION. Minas Gerais.
SPECIMENS EXAMINED. MINAS GERAIS: Mun. de Santana do Riacho, 116 km ao longo da rodovia Belo Horizonte – Conceição do Mato Dentro, 19 April 1981, *Rossi & Amaral* SPF 19995 (K).
HABITAT. Insufficient data.
CONSERVATION STATUS. Data Deficient.

Superficially similar to *Stachytarpheta glabra*, but with bigger leaves, and with a sparse covering of hairs on stems, leaves, calyx etc., and with a longer inflorescence. Schauer (1851) cites only the Sellow specimen, which he saw in Berlin. This was presumably the holotype on which Chamisso based the species, and would have been destroyed in Berlin. The *Sellow* specimen at Kew is a good match for the description, but it bears a number, 1456, and it is hard to understand why this number was not cited, either by Chamisso or Schauer. There is no further locality or information on this specimen.

71. **Stachytarpheta cassiae** *S. Atkins* **sp. nov.** a *S. glabra* foliis late ovatis (non ellipticis usque late ellipticis), inflorescentiis usque 4 cm longis (non usque 7 cm), bracteis ubique glabris (non margine ciliatis) differt. Typus: Minas Gerais, Serra do Pau Dárco, 15 March 1994, *Sakuragui et al.* CFCR 15086 (holotypus SPF; isotypus K).

Rounded shrub to 2 m, much-branched. Stems woody, 4-sided, glabrous to sparsely covered with short stiff hairs. Leaves petiolate, patent to erect, thickly chartaceous, broadly ovate, (2 –)3.5 – 4.5 × (1.5 –)3 – 3.8 cm, apex obtuse to rounded, base attenuate, decurrent into petiole, margin crenate, slightly inrolled, upper surface glabrous except for very few scattered hairs along main vein, occasionally with scattered large nectaries, lower surface glabrous, covered with sessile glands. Inflorescence 3 – 4 cm by 15 mm; bracts linear, c. 10 mm long. Calyx 12 mm long, 4-toothed abaxially, with outer teeth longer than inner, sinus adaxial, outer surface scattered with sessile glands. Corolla blue, hypocrateriform, tube straight, c. 20 mm long, outer surface with very short glandular hairs, lobes c. 1 mm across. Stamens inserted above middle of tube. Ovary flask-shaped, c.

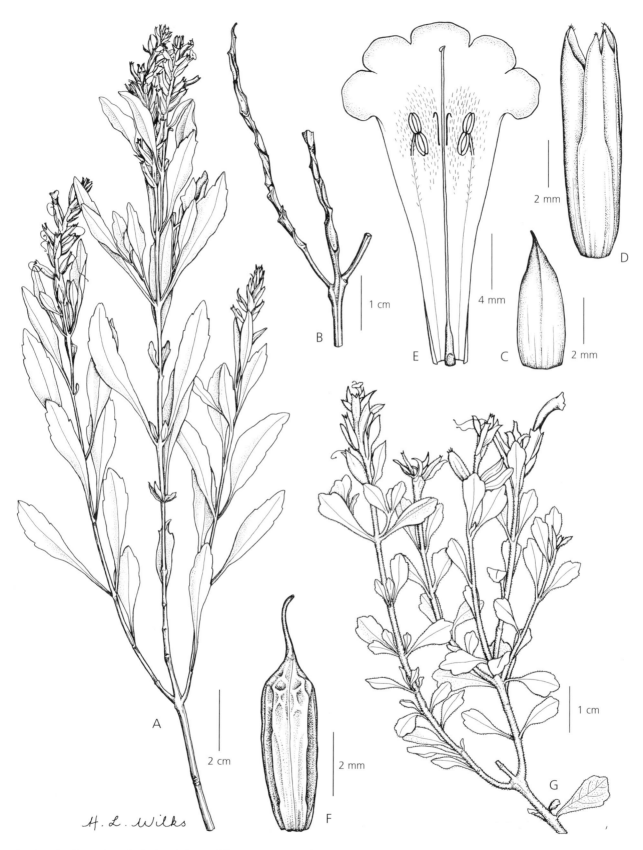

Fig. 30. *Stachytarpheta harleyi.* **A** habit; **B** inflorescence rachis from previous year; **C** bract; **D** calyx; **E** corolla; **F** fruit. *Stachytarpheta diamantinensis.* **G** habit. **A** from *Hatschbach et al.* 66320; **B** from *Hatschbach et al.* 72073; **C – F** from *França et al.* 4543; **G** from *Anderson et al.* 35436. DRAWN BY HAZEL WILKS.

2 mm long, style c. 20 mm long. Fruit dark brown, surface reticulate, without prominent attachment scar, with stylopodium. Fig. 29A – E, Plate 1C.

DISTRIBUTION. Minas Gerais (Map 24).
SPECIMENS EXAMINED. MINAS GERAIS: Entre os Mun. de Espinosa e Montezuma, Serra do Pau Dárco, 15 March 1994, *Roque et al.* CFCR 15052 (K).
HABITAT. *Campo rupestre*, at 1300 – 1400 m.
CONSERVATION STATUS. Vulnerable. Narrow distribution. Potentially threatened due to its limited range and generally small size of its known populations.

A much more bushy and rounded shrub than *Stacytarpheta glabra*, but almost completely glabrous, and with a similar lax inflorescence. This species is named for Cassia Sakuragui who collected the type specimen.

72. Stachytarpheta diamantinensis *Moldenke* (1973a: 220). Type: Minas Gerais, 5 km SW of Diamantina, 5 Feb. 1972, *Anderson et al.* 35436 (holotype TEX; isotypes NY, K).

Shrub to 50 cm, branched. Stem rounded, quite densely covered with short, uniseriate hairs, some dark-coloured, possibly viscid. Leaves petiolate, often with smaller leaves in the same axil, patent, narrowly ovate, 6 – 14 × 2 – 6 mm, apex acute, base attenuate, decurrent into petiole, margin slightly inrolled, shallowly 3-lobed at apex, upper and lower surface tomentose with short, uniseriate hairs, more dense on lower surface. Inflorescence 2 – 4 cm by c. 10 mm; bracts linear, c. 10 mm long, covered with short, uniseriate hairs. Calyx c. 11 mm long, with 4 teeth abaxially, outer teeth longer than inner, sinus adaxially, covered with short, uniseriate hairs. Corolla blue, hypocrateriform, tube straight, lobes small. Fruit not seen. Fig. 30G.

Only known from the type.
HABITAT. Open hillside on sandy soil with sandstone boulders at 1300 m.
CONSERVATION STATUS. Vulnerable. Only a single known locality. This area is high in endemics and there is currently no protection. This taxon is at risk.

This plant differs from the rest of this group by its smaller leaves with covering of short hairs, and by its shorter inflorescence.

Group 11 Martiana
Two species in Brazil.
The species in this group are similar to those in the Glabra group, but they differ in the calyx, bracts and fruit.

Key to species in the Martiana group

1 Leaves ovate, 2 –5 × 1 – 2 cm; bracts c. 4 mm long, broadly triangular, glabrous except for margin · 74. **S. hatschbachii**
   Leaves obovate to ovate, 3.5 – 13 × 1.5 – 3.5 cm; bracts 7 mm long, ovate, acuminate, covered with uniseriate hairs and large nectaries · 73. **S. martiana**

73. Stachytarpheta martiana *Schauer* (1847: 568; 1851: 212). Type: Minas Gerais, in campis editis, *Martius* s.n. (holotype M n.v.; photo NY). (In *Fl. Bras.* Schauer gives the locality as: 'in campis altis ad fluvium Itacambirussu').
S. obovata Hayek (1907: 273). Type: Brazil, Goiás, *Gardner* 4337 (isotypes E, NY, K) .

Shrub to 1.5 m, branched. Stems woody, rounded, glabrous, pale. Leaves petiolate, patent to erect, thick, obovate to ovate, 3.5 – 7 × 1.5 – 2.8 cm, apex obtuse, base attenuate, decurrent into petiole, margin crenate, upper surface almost glabrous except for hairs along main veins, lower surface sparsely covered with short white hairs, large glands also present (randomly). Inflorescence 16 cm long × 12 mm wide; bracts 7 mm long, ovate, acuminate, covered with uniseriate hairs and large nectaries. Calyx c. 11 mm long, 4-toothed with sinus on adaxial side, outer surface covered with minute hairs and nectaries. Corolla dark blue, hypocrateriform, tube slightly bent, 17 mm long, lobes c. 6 mm across, throat hairy. Stamens lying at top of tube. Ovary ovoid. Fruit dark brown, c. 6 mm long with prominent attachment scar.

DISTRIBUTION. Goiás & Minas Gerais (Map 25).
SPECIMENS EXAMINED. GOIÁS: Between São Domingos and Posse, 1840, *Gardner* 4337 ; Mun. de Posse, na estrada para Guarani de Goiás, 27 April 1996, *Pereira*

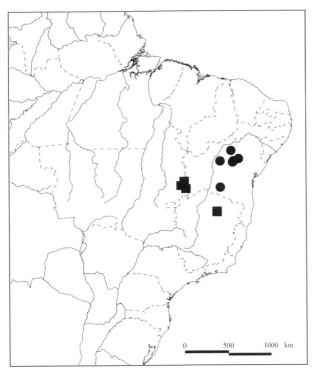

Map 25. Distributions of *Stachytarpheta hatschbachii* (●) and *S. martiana* (■).

& Alvarenga 2992 (K). **MINAS GERAIS:** Grão Mogol, margem da estrada Grão Mogol – Montes Claros, 23 May 1982, *Hensold et al.* CFCR 3504/SPF 30083 (K); Grão Mogol – Cristália, 15 April 1981, *Cordeiro et al.* CFCR 958/SPF 22885 (K); Grão Mogol, estrada para Francisco Sá, 15 Oct. 1988, *Harley et al.* 25058 (K).

HABITAT. *Cerrado/campo rupestre*, at 700 – 950 m.

CONSERVATION STATUS. Least concern.

One of the few species with a distribution on both sides of the São Francisco river, although quite restricted at each locality.

A largish shrub forming thickets. Quite similar to *Stachytarpheta hatschbachii* from Bahia, but with larger leaves.

74. Stachytarpheta hatschbachii *Moldenke* (1980: 39). Type: Bahia, Mun. Morro do Chapéu, Serra do Tombador, 15 July 1979, *Hatschbach & Guimarães* 42347 (holotype TEX; isotype MBM).

Shrub to 1.5 m. Stems woody, 4-sided, somewhat gnarled, with scattering of short white hairs. Leaves petiolate, patent, often conduplicate, ovate, fleshy, 2 – 5 × 1 – 2 cm, apex obtuse, base attenuate, margin crenate, upper and lower surfaces almost glabrous with a scattering of short, white hairs, and many dark sessile glands covering surface. Inflorescence 6 – 12 cm long × 10 mm wide; bracts ovate, glabrous except for margin, c. 6 mm long. Calyx c. 12 mm long, glabrous, 4-toothed with sinus on adaxial side. Corolla intense blue, hypocrateriform, tube straight, c. 20 mm long, lobes c. 4 mm across. Stamens inserted towards top of tube. Ovary asymmetrical, 3 mm long; style exserted c. 2 cm long. Fruit dark brown, reticulate, with stylopodium, and prominent attachment scar, and with distinct ridge between mericarps. Fig. 31, Plate 5A.

DISTRIBUTION. Bahia (Map 25).

SPECIMENS EXAMINED. BAHIA: Jacobina, 1840, *Blanchet* 3126 (K); Serra do Curral Feio, 6 March 1974, *Harley et al.* 16840 (K); Morro do Chapéu, 26 Aug. 1981, *Gonçalves* 127 (K); 28 June 1983, *Coradin et al.* 6211 (K); 18 Nov. 1986, *Webster et al.* 25752 (NY; K); Estrada Morro do Chapéu – Jacobina, 29 June 1996, *Harley et al.* PCD 3277 (K); Mun. Caetité, Barragem de Carnaíba, 10 Feb. 1997, *Passos et al.* PCD 5399 (K); Xique – Xique, Mun. Gentio do Ouro, 13 March 1998, *Hatschbach et al.* 67727 (MBM; K); Casa Nova, 5 July 2003, *de Queiroz et al.* 7894 (HUEFS).

HABITAT. *Campo rupestre* and *caatinga*, in sandy soil at 900 – 1100 m.

CONSERVATION STATUS. Vulnerable. This taxon occurs at the extreme north of the Espinhaço range. The *campo rupestre* here is fragmented, and there is no protection, although this is a site of high endemism. The area is under pressure from encroaching agriculture and this taxon could be at risk.

This species has a very pronounced commissure giving a strong vertical ridge to the fruit.

Group 12 Caatingensis

Two species in Brazil.

The two species in this group differ from all others in the genus because of the distant arrangement of flowers on the inflorescence rachis, and the angle of the calyx from the rachis (c. 45° – c. 90°).

Key to species in the Caatingensis group

1. Inflorescence to 27 cm; calyx and bracts fairly densely covered with a mixture of short, gland-topped and long uniseriate hairs ·································· **75. S. caatingensis**

   Inflorescence to 13 cm; calyx and bracts only sparsely covered with a mixture of short, gland-topped and long uniseriate hairs ·································· **76. S. brasiliensis**

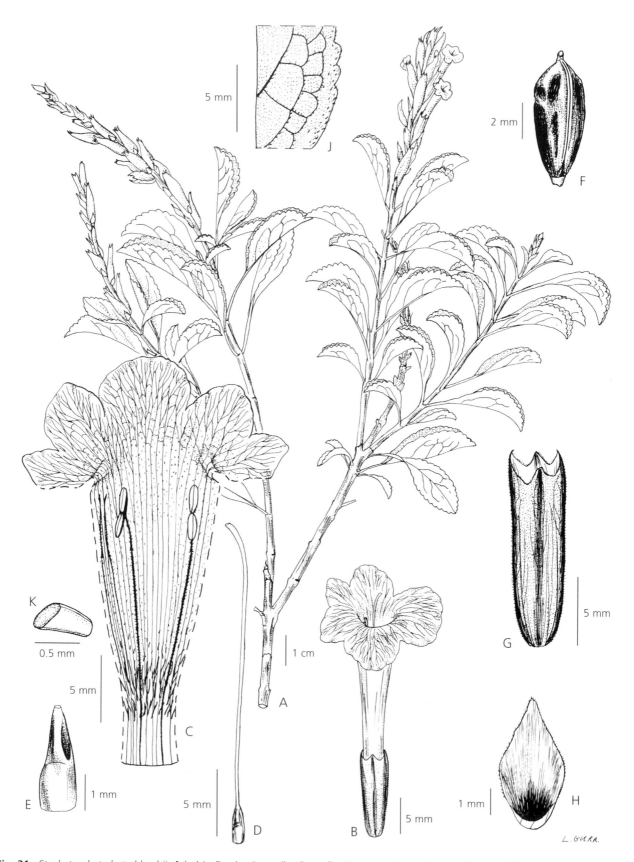

Fig. 31. *Stachytarpheta hatschbachii*. **A** habit; **B** calyx & corolla; **C** corolla; **D** ovary, style and stigma; **E** ovary; **F** fruit; **G** calyx; **H** bract; **J** detail of leaf undersurface; **K** apex of style. **A, E – K** from *Harley et al.* 16840; **B – D** from *Harley et al.* PCD 3277. DRAWN BY LINDA GURR.

75. Stachytarpheta caatingensis *S. Atkins* **sp. nov.** a *S. brasiliensi* inflorescentiis longioribus (usque 27 cm, non 13 cm), calycibus bracteisque moderate dense (non sparse) pilis brevibus glandulis terminantibus et pilis longis uniseriatis intermixtis obsitis differt. Typus: Bahia, Tanhaçu, 11 Jan. 2000, *Giulietti & Harley* 1701 (holotypus HUEFS; isotypus K).

Stachytarpheta villosa var. *bahiensis* Moldenke (1983b: 400). Type: Bahia, Aracatu, 15 May 1983, *Hatschbach* 46386 (holotype TEX; isotype NY, ?MBM n.v.).

Shrub to 1.5 m, branching below inflorescence. Stems woody, rounded, glabrous to sparsely hairy with uniseriate hairs interspersed with short gland-topped hairs. Leaves short-petiolate, patent, chartaceous, ovate to narrowly ovate, 2 – 7 × 0.8 – 2.5 cm, apex acute, base attenuate, decurrent into petiole, margin crenate, upper and lower surfaces sparsely to fairly densely covered with uniseriate hairs interspersed with short gland-topped hairs. Inflorescence 7 – 12 × 2 cm, rachis narrow, c. 1 mm wide, with calyces held at 45 – 90° angles, widely spaced; bracts subulate c. 5 mm long, covered with short gland-topped hairs. Calyx c. 10 mm long, 4-toothed, teeth ± equal, with adaxial sinus, outer surface covered with gland-topped hairs. Corolla blue, hypocrateriform, tube straight, c. 11 mm long, lobes c. 6 mm across, throat glandular. Stamens inserted towards top of tube, included. Ovary oblong, c. 2 mm long, style 9 mm long. Fruit light brown, surface smooth. Fig. 32, Plate 5B.

DISTRIBUTION. Bahia (Map 26).
SPECIMENS EXAMINED. BAHIA: Contendas do Sincorá, 22 Nov. 1985, *Hatschbach & Zelma* 50080 (K); Mun. de Aracatu, rod. de Brumado para Vitória da Conquista, 29 Dec. 1989, *Carvalho et al.* 2698 (K); Anajé, 27 Jan. 1965, *Pereira & Pabst* 8657 (K); Estação Ecologica do Raso da Catarina, 24 June 1984, *de Queiroz* 296 (K); 1 – 2 km SE de Tanhaçu, 22 Jan. 1997, *Arbo et al.* 7665 (NY; K); Tanhaçu, estrada para Ourives, 1 Feb. 2003, *França et al.* 4188 (HUEFS; K).
HABITAT. *Caatinga.*
CONSERVATION STATUS. Vulnerable. The distribution of this species is fragmented and only one of the localities is protected.

76. Stachytarpheta brasiliensis *Moldenke* (1934: 239). Type: Minas Gerais, Serra do Caraça, Oct. 1882, *Glaziou* 14161 (holotype P n.v.; isotype K).

Subshrub. Stems 4-sided, sparsely to densely covered with uniseriate hairs. Leaves petiolate, ovate, 3.5 – 8.5 × 1.5 – 3 cm, apex acute, base attenuate, decurrent into petiole, crenate, upper and lower surface sparsely covered with simple hairs, more densely so along veins. Inflorescence 8 – 13 × c. 1 cm, rachis very thin hardly excavated, flowers distant, alternate, and patent to rachis; bracts ovate, rostrate at apex, with uniseriate hairs and some gland-tipped hairs scattered along margin, c. 5 mm long. Calyx ribbed with scattered gland-tipped hairs.

No modern material to complete description.
DISTRIBUTION. Minas Gerais (Map 26).
SPECIMEN EXAMINED. MINAS GERAIS: Serra do Caraça, *Glaziou* 13061 (K).
HABITAT. Only Glaziou's notes for information " dans le bois et le campo."
CONSERVATION STATUS. Data deficient. The Caraça Massif is isolated from other parts of the Serra do Espinhaço range by areas of lower elevation, formerly covered in high forest. Some parts of the Serra do Caraça are protected by the Reserva Particular de Patrimônio Natural do Caraça.

Although similar morphologically, the two species in this group are from very different habitats. *Stachytarpheta caatingensis* grows in dry *caatinga* at low altitudes, and *S. brasiliensis* is from the Serra do Caraça which is an area of relatively high rainfall with forest at lower altitudes and well-developed *campo rupestre* at higher altitudes.

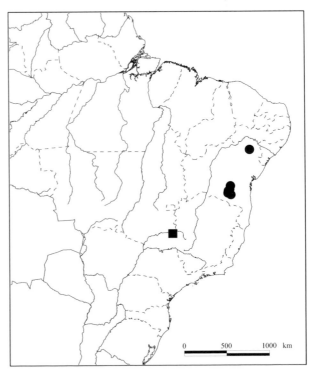

Map 26. Distributions of *Stachytarpheta brasiliensis* (■) and *S. caatingensis* (●).

KEW BULLETIN VOL. 60(2)

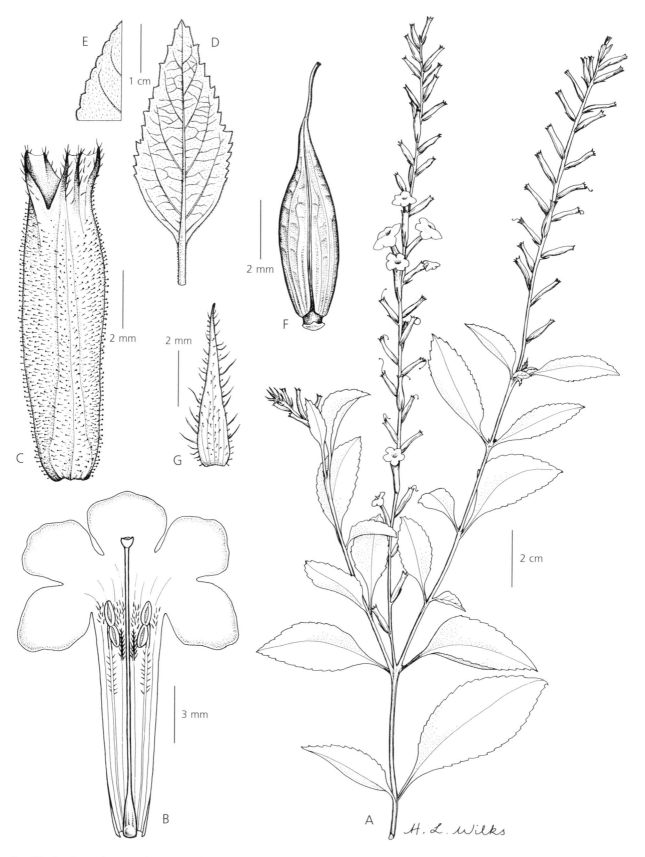

Fig. 32. *Stachytarpheta caatingensis.* **A** habit; **B** corolla; **C** calyx; **D** leaf undersurface; **E** detail of leaf undersurface; **F** fruit; **G** bract. **A – C** and **F – G** from *Giulietti & Harley* 1701; **D & E** from *Pereira & Pabst* 8657. DRAWN BY HAZEL WILKS.

Incertae sedis

77. Stachytarpheta pachystachya *Mart. ex Schauer* (1847: 569; 1851: 214); Dawson (1957). Type: "campis altis prov. Minarum, *Martius* s.n., *Claussen* s.n., *Riedel* s.n., Piauhy [Piauí] et Goyaz [Goiás]: Serra do Duro, 1839, *Gardner* 3410 (isosyntype K).

Perennial herb/subshrub to 1.5 m. Stems square, densely covered in long, uniseriate hairs, more dense at nodes. Leaves sessile to petiolate, patent, with many smaller leaves in the axils, chartaceous, elliptic, 2 – 5 × 1 – 2 cm, apex acute, base acute, decurrent into petiole, margin serrate-crenate, upper and lower surface quite densely covered with uniseriate hairs and sessile glands. Inflorescence 6 – 22 cm long × 20 mm wide; bracts ovate, apex long-acuminate, base truncate, sparsely hairy and glandular, margins with uniseriate hairs. Calyx c. 11 mm long, 5-lobed with abaxial sinus, outer surface with uniseriate hairs. Corolla blue-purple, tube straight, c. 12 mm long, hairy and sparsely glandular at throat, lobes small, c. 2 mm across. Stamens ± sessile, inserted just below throat. Ovary flask-shaped, glabrous. Fruit black, surface reticulate, shiny, without prominent attachment scar, rounded at apex. Fig. 33.

DISTRIBUTION. Goiás, Minas Gerais and Piauí. Minas Gerais localities may be dubious.

SPECIMENS EXAMINED. GOIÁS: 1840, *Gardner*, 3935 (K); Chapada dos Veadeiros, 20 km N of São João da Aliança, 14 April 1956, *Yale Dawson* 14209 (RSA); 24 km by road SW of Monte Alegre de Goiás, 11 March 1973, *Anderson et al.* 6808 (K); 13 km by road S of São João da Aliança, 21 March 1973, *Anderson et al.* 7555 (K); Mun. São Domingos, estrada para Posse, 24 Feb. 2003, *França et al.* 4654. **MINAS GERAIS:** *Claussen* sn. (K); 1892, *Glaziou* 19718 (K).

HABITAT. *Cerrado.*

CONSERVATION STATUS. Data deficient. One of the few species occurring on both sides of the São Francisco river, although quite restricted at each locality. Not very frequently collected, and collections close to habitation.

This species is close to those in Group 2 Gesnerioides, in that it has the fruit tip slightly flattened and elongated, but it differs from the rest of the group by the inflorescence being of medium length, and quite broad. It is the only species to have a tightly-packed inflorescence, 2 – 3 flowers broad, up to 22 cm long, and with broadly ovate bracts.

78. Stachytarpheta glandulosa *S. Atkins* **sp. nov.** *S. quadrangulae* similis sed calyce 4-lobato (non 2), fructu reticulato (non laevi), petiolis multiglandulosis (non eglandulosis) distincta. Typus: Bahia, Licínio de Almeida, c. 12 km da cidade em direcao a Brejinho das Ametistas, 12 March 1994, *Roque et al.* CFCR 15014 (holotypus SPF; isotypus K).

Shrub to 1.7 m. Stem rounded to 4-sided, dark, somewhat gnarled. Leaves petiolate, thick, ovate, 2.5 – 6 × 1.4 – 3.8 cm, apex obtuse, base truncate, margin closely crenate, upper surface fairly densely covered with very short uniseriate hairs, the lower surface much more densely covered, both surfaces scattered with small, white glands, especially numerous on the petioles. Inflorescence 7 – 11 cm long × 4 – 5 mm wide; bracts short-triangular, 2 – 4 mm long. Calyx 10 mm long, 4-toothed with sinus, fairly densely covered with very short hairs. Corolla hypocrateriform, violet-blue, tube 22 mm long, lobes c. 3 mm across. Stamens lying just below throat. Ovary flask-shaped, c. 2 mm long. Fruit dark brown, slightly reticulate, beaked with short stylopodium. Fig. 34.

DISTRIBUTION. Bahia

SPECIMENS EXAMINED. Only known from the type.

HABITAT. *Campo rupestre* at 800 – 900 m.

CONSERVATION STATUS. Data deficient. Only one collection.

This species is difficult to place. It has a very thin rachis, with the flowers rather lax, and the base of the calyx inserted. However, the corollas are extremely long, the calyx is 4-toothed and the fruit is beaked and reticulate.

79. Stachytarpheta andersonii *Moldenke* (1974a: 303). Type: Goiás, 2 – 4 km N of Funil and the rio Parana at 600 m, *cerrado* and grassy *campo*, 14 March 1973, *Anderson* 7105 (UB; fragment & photo NY). Known only from unicate material in Brasilia, plus a fragment in NY. Undoubtedly a good species, but no material is available to allow a new description.

Doubtful taxa

Stachytarpheta angustifolia forma **rionigrensis** *Moldenke* (1977: 408). Type: Habitat in sylvis ad M. araracoara, Prov. Rio Negro, Febr. *Martius*, Iter Brasil (holotype ? B†; fragment TEX). It is not possible from the fragment to confirm this identification.

Stachytarpheta gesnerioides var. **simplex** (*Hayek*) *Moldenke* (1974c: 193). Type: Brasilia orientalis, *Tamberlik* s.n. (Type as for *S. simplex* Hayek — see below). Type not seen.

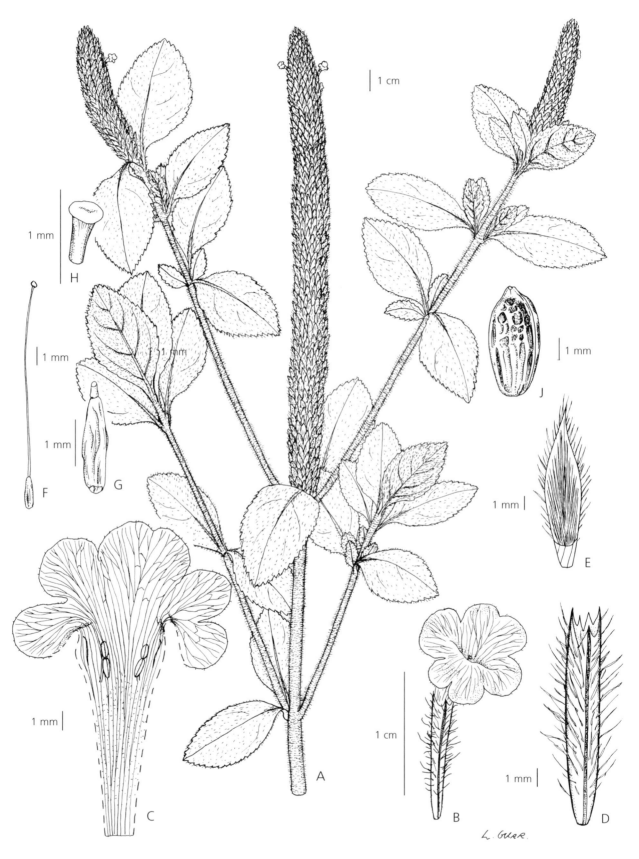

Fig. 33. *Stachytarpheta pachystachya.* **A** habit; **B** flower; **C** corolla; **D** calyx; **E** bract; **F** gynoecium; **G** ovary; **H** stigma; **J** fruit. **A** from Gardner 3935; **B – H** from *Hatschbach et al.* 71069; **J** from *Anderson* 7555. DRAWN BY LINDA GURR.

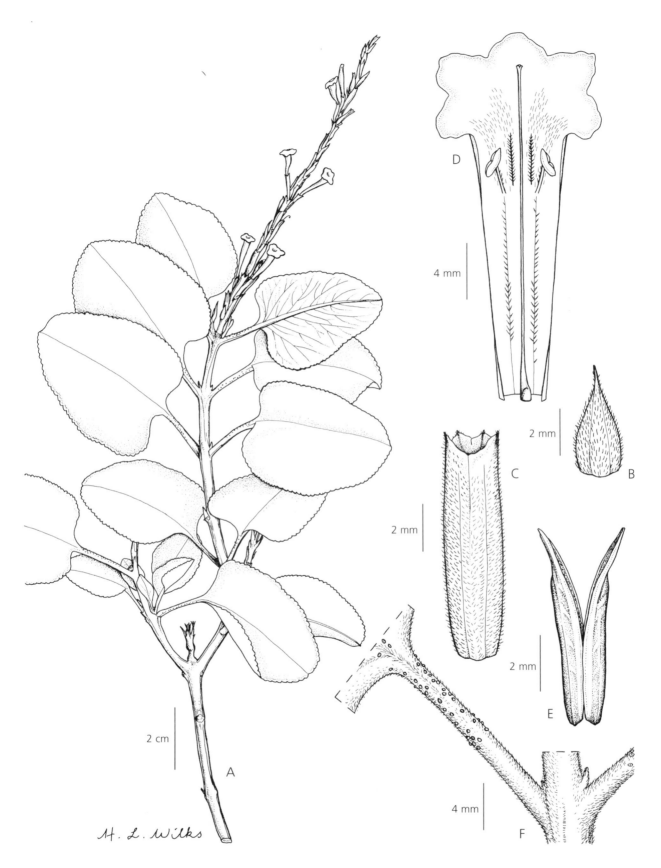

Fig. 34. *Stachytarpheta glandulosa.* **A** habit; **B** bract; **C** calyx; **D** corolla; **E** fruit; **F** detail of petiole showing glands. From *Roque et al.* CFCR 15104. DRAWN BY HAZEL WILKS.

Stachytarpheta lanata *Schauer* (1847: 571). Type: In campis generalibus prov. Minas, *Martius* s.n. (holotype M n.v.; isotype BR). Not enough material to interpret the type. No other material.

Stachytarpheta maximiliani var. **glabrata** *Schauer* (1847: 565). Type: *Gardner* s.n. (?holotype G). Schauer (1847) made an informal variety based on a *Gardner* specimen without number. In G is a glabrous form, noted as such, perhaps by Schauer himself, with "Gardner 1105 ou 1115" on the label. It is not possible to know for certain if this is the specimen he based his variety on. It appears no more glabrous than other specimens. I remain uncertain about the status of this variety.

Stachytarpheta rotundifolia *Link.* Walpers put this into the synonymy of *S. hirsutissima* Link. The type cited by Walpers is: Link in Herb. Reg. Berolin - crescit in Brasilia. (v.s. cult.). I have not been able to establish a type specimen to confirm this taxon (Type ?B†).

Stachytarpheta simplex *Hayek* (1907: 273). Type: Brasilia orientalis, *Tamberlik* s.n. (Type not seen).

Stachytarpheta subulata *Moldenke* (1940: 475). Type: Bahia, *Blanchet* 3139a, (F). Immature specimen, most probably *S. cayennnensis*.

Unidentified specimens

Mueller 146 (K) — Santa Catherina. This could be a taxon as yet undescribed, but not enough material or information.
Gardner 4338 (K) — Tocantins/Goias, Arraias. An undescribed taxon, but not enough material or information.
Bernacci & Bernacci 1426 (SP; K) - São Paulo, Serra da Agua Limpa, *cerrado* & *mata decidua*. I cannot place this. Could be undescribed, but not enough material.

Acknowledgements

I wish to thank the Directors of BR, E, F, G, HUEFS, NY, RSA, S and TEX for the generous loan of material for this paper. I also thank Ray Harley for the original inspiration to start this research, and subsequent help and support, including help with the final draft of the manuscript. He has also made his photographs available. I would like to thank Dr Alan Paton for providing help and support with the cladistics, and during the final stages, including reading the manuscript. I thank Drs Daniela Zappi, Nigel Taylor and Mark Coode for their comments on the final draft of the paper, and Mark Coode for making the latin diagnoses and for providing one of the photographs, Justin Moat for producing the maps, Dr Dick Brummitt for help with nomenclatural problems, and Margaret Tebbs, Linda Gurr and Hazel Wilks for their illustrations. I thank my Brazilian colleagues who have given me help in the field, including Drs Paulo Sano, Cassia Sakuragai, Vinicius Souza, Tânia da Silva, Fatima Salimena, Lenise Guedes, Eduardo Saar, Luzinaldo Passos and on the most recent trip I owe a great deal to Dra Efigênia Melo, Benedito Marques da Silva, and not least, my colleague Dr Flávio França who was able to be my 'responsavel'. Flávio has also helped with information on collection localities. I also thank Dra Ana Maria Giulietti for enthusiatic encouragement during the whole project. Thanks to the Kew editorial staff, Dr M. Lock and Ruth Linklater for improvement of the manuscript and Christine Beard for typesetting. I thank Dr Jovita Yesilyurt for the Portuguese translation of the summary and Dr David Simpson for helpful amendments to the text. I also thank British Airways who were able to support some of the travel costs through their *Community and Conservation* scheme, which has aided Kew botanists since 1992.

References

Atkins, S. (1991). *Stachytarpheta sericea* Atkins (*Verbenaceae*) and its hybrid with *S. chamissonis* Walp. Kew Bull. 46: 281 – 289.
—— (2004). *Verbenaceae*. In: K. Kubitzki & J. W. Kadereit (eds.), The Families and Genera of Vascular Plants, Vol. VII: 449 – 468. Springer-Verlag, Germany.
——, Alves, R. J. V. & Kolbeck, J. (1996). Plants in Peril, 23 — *Stachytarpheta sellowiana*. Curtis's Bot. Mag. 13: 33 – 35.
Baillon, H. E. (1892). Histoire des plantes 11. Labiées – Ilicacées. L. Hachette et Cie, Paris.
Briquet, J. (1895). *Verbenaceae*. In: H. G. A. Engler & K. Prantl, Die natürlichen Pflanzenfamilien IV, 3a: 132 – 182. W. Engelmann, Leipzig.
Chamisso, L. K. A. von (1832). De plantis in expeditione Romanzoffiana et in Herbariis Regiis, *Verbenaceae*. Linnaea 7: 213 – 272.
Danser, B. H. (1929). Über die Niederlandische-Indischen *Stachytarpheta*-Arten und ihre Bastarde, nebst Betrachtungen über die Begrenzung der Arten im Allgemeinen. Ann. Jard. Bot. Buitenzorg Vol. 40: 1 – 44.
Dawson, E. Y. (1957). The Machris Brazilian expedition: botany: Phanerogamae, various smaller families. Los Angeles County Mus. Contr. Sci. 7.
Dubs, B. (1998). Prodromus Florae Matogrossensis. Betrona-Verlag, Switzerland.
Felsenstein, J. (1985). Confidence limits on phylogenies: an approach using the bootstrap. Evolution 39: 783 – 783.

Fernandes, R. (1984). Notes sur les *Verbenaceae*: 1. Remarques sur quelques espèces de *Stachytarpheta* Vahl. Bol. Soc. Brot. 57: 87 – 111.

Filgueiras, T. S. (1997). Distrito Federal. In: S. D. Davis, V. H. Heywood, O. Herrera-MacBryde, J. Villa-Lobos & A. C. Hamilton (eds.), Centres of Plant Diversity. A Guide and Strategy for their conservation. Vol. III The Americas: 405 – 410. WWF & IUCN, Cambridge.

Fitch, W. M. (1971). Toward defining the course of evolution: minimum change for a specific tree topology. Syst. Zool. 20: 406.

Galvão, M. V. & Nimer, E. (1965). Clima in Geografia do Brasil — Grande Região Leste, IBGE, Rio de Janeiro, 5 (19): 91 – 139.

Gentry, A. H. (1982). Patterns of neotropical plant species diversity. In: M. K. Hecht, B. Wallace & G. T. Prance (eds.), Evolutionary Biology 15: 1 – 84.

Giulietti, A. M. & Pirani, J. R. (1988). Patterns of geographic distribution of some plant species from the Espinhaço Range, Minas Gerais and Bahia, Brazil. In: P. E. Vanzolini & W. R. Heyer (eds.), Proceedings of a workshop on neotropical distribution patterns: 39 – 69. Academia Brasileira de Ciências, Rio de Janeiro.

—— *et al.* (1987). Flora da Serra do Cipó, Minas Gerais. Caracterização e lista das especies. (Flora of the Serra do Cipó, Minas Gerais: characterization and check-list of species). Bol. Bot. (São Paulo) 9: 1 – 151.

—— & Pirani, J. F. (1997). Espinhaço Range Region, Eastern Brazil. In: S. D. Davis, V. H. Heywood, O. Herrera-MacBryde, J. Villa-Lobos & A. C. Hamilton (eds.), Centres of Plant Diversity. A Guide and Strategy for their Conservation. Vol. III The Americas: 397 – 404. IUCN.

Glaziou, A. F. M. (1911). Liste des plantes du Brésil Central recueillies en 1861 – 1895. Mem. Soc. Bot. France 1 (3): 544.

Greuter, W. *et al.* (eds.). (2000). International Code of Botanical Nomenclature (Saint Louis Code). Koeltz Scientific Books, Königstein.

Harley, R. M. (1986). Observations on a hybrid population of *Hyptis cruciformis* and *H. pachyphylla* in Brazil. Notes on New World *Labiatae* X. Kew Bull. 41: 1007 – 1015.

—— (1988). Evolution and distribution of *Eriope* (*Labiatae*) and its relatives in Brazil. In: P. E. Vanzolini & W. R. Heyer (eds.), Proceedings of a workshop on neotropical distribution patterns: 71 – 120. Academia Brasileira de Ciências, Rio de Janeiro.

—— (1995). Introduction. In: B. L. Stannard (ed.), Flora of the Pico das Almas, Chapada Diamantina, Bahia, Brazil. Royal Botanic Gardens, Kew.

—— & Simmons, N. A. (1986). Florula of Mucugê, Chapada Diamantina, Bahia, Brazil. A descriptive check-list of the *campo rupestre* area. Royal Botanic Gardens, Kew.

Hayek, A. (1907). *Verbenaceae* novae herbarii Vindobonensis. Repert. Spec. Nov. Regni Veg. 3: 273 – 274.

Hooker, J. D. (1865). *Stachytarpheta bicolor.* Bot. Mag: t. 5538.

Humboldt, A., Bonpland, A. & Kunth, C. S. (1815). Nova genera et species plantarum II. Paris.

IUCN (2001). IUCN Red List categories and criteria, version 3.1. IUCN, Gland, Switzerland & Cambridge, U.K.

Jacquin, J. F. (1844). Eclogae Plantarum Rariorum 2. Antonii Strauss, Vienna.

Jacquin, N. J. von (1771). Observationum Botanicarum 4: 7, t. 86.

Jansen-Jacobs, M. J. (1988). *Verbenaceae.* In: A. R. A. Görts-van Rijn (ed.), Flora of the Guianas. Koeltz, Germany.

Junell, S. (1934). Zur Gynaceummorphologie und Systematik der Verbenaceen und Labiaten. Symb. Bot. Upsal. 4: 1 – 219.

Link, H. F. (1821). Enumeratio Plantarum Horti regii botanici berolinensis altera. Berlin.

Linnaeus, C. (1753). Species Plantarum 1. Stockholm.

—— (1759). Systema Naturae Ed. 10, 2: 851. Stockholm.

—— (1762). Species Plantarum Ed. 2: 27. Stockholm.

Mansfeld, R. (1924). *Verbenaceae.* In: R. Pilger, Plantae Luetzelburgianae brasilienses V. Notizbl. Bot. Gart. Berlin 9: 153 – 156.

Medikus, F. K. (1789). Philosophische Botanik. Hofund Akademischen Buchhandlumg, Mannheim.

Miller, P. (1768). The gardeners dictionary, ed. 8. London.

Moldenke, H. N. (1934). Some new and neglected species and varieties of the *Verbenaceae.* Repert. Spec. Nov. Regni Veg. 37: 209 – 239.

—— (1940). Novelties among the American *Verbenaceae.* Phytologia 1: 453 – 480.

—— (1941a). Novelties in the *Eriocaulaceae* and *Verbenaceae.* Phytologia 2: 6 – 32.

—— (1941b). Plant novelties. Phytologia 2: 50 – 57.

—— (1947a). Notes on new and noteworthy plants II. Phytologia 2: 306 – 324.

—— (1947b). Notes on new and noteworthy plants III. Phytologia 2: 363 – 372.

—— (1947c). Notes on new and noteworthy plants I. Phytologia 2: 213 – 242.

—— (1948). Notes on new and noteworthy plants V. Phytologia 2: 464 – 477.

—— (1950a). Notes on new and noteworthy plants X. Phytologia 3: 261 – 281.

—— (1950b). Notes on new and noteworthy plants XI. Phytologia 3: 307 – 320.

—— (1956). Notes on new and noteworthy plants. Revista Sudamer. Bot. 10 (7): 229 – 232.

—— (1958). Notes on new and noteworthy plants XXII. Phytologia 6: 322 – 332.

—— (1959a). A resumé of the *Verbenaceae*, *Avicenniaceae*, *Stilbaceae*, *Symphoremaceae* and *Eriocaulaceae* of the world as to valid taxa, geographic distribution and synonymy. New Jersey.

—— (1959b). Notes on new and noteworthy plants XXIV. Phytologia 7: 77 – 91.

—— (1966). Notes on new and noteworthy plants XLV. Phytologia 13: 306 – 307.

—— (1968). A new species of *Stachytarpheta* from Brazil. Phytologia 16: 487.

—— (1970). Taxonomic notes on the *Eriocaulaceae* and *Verbenaceae*. Phytologia 20: 242 – 243.

—— (1971). A fifth summary of the *Verbenaceae*, *Avicenniaceae*, *Stilbaceae*, *Dicrastylidaceae*, *Symphoremaceae*, *Nyctanthaceae* and *Eriocaulaceae* of the world as to valid taxa, geographic distribution, and synonymy. 2 Vols. New Jersey.

—— (1972). Two verbenaceous novelties. Phytologia 24: 454 – 455.

—— (1973a). Notes on new and noteworthy plants LVI. Phytologia 25: 220 – 224.

—— (1973b). Seven novelties from North and South America. Phytologia 25: 117 – 120.

—— (1974a). Notes on new and noteworthy plants LXVIII. Phytologia 28: 303 – 304.

—— (1974b). Notes on new and noteworthy plants LXX. Phytologia 28: 466 – 468.

—— (1974c). Notes on new and noteworthy plants LXXII. Phytologia 29: 192 – 193

—— (1974d). Notes on new and noteworthy plants LXXI. Phytologia 29: 75 – 78.

—— (1975). Notes on new and noteworthy plants LXXX. Phytologia 31: 373.

—— (1976). Notes on new and noteworthy plants LXXXVII. Phytologia 33: 372 – 375.

—— (1977). Notes on new and noteworthy plants CV. Phytologia 37: 408.

—— (1978). Notes on new and noteworthy plants CXIV. Phytologia 40: 54.

—— (1979a). Notes on new and noteworthy plants CXXI. Phytologia 41: 449 – 451.

—— (1979b). Notes on new and noteworthy plants CXXXII. Phytologia 44: 473.

—— (1980). Notes on new and noteworthy plants CXXXIII. Phytologia 45: 36 – 40.

—— (1981a). Notes on new and noteworthy plants CXLV. Phytologia 48: 253 – 255.

—— (1981b). Notes on new and noteworthy plants CXLIV. Phytologia 47: 330.

—— (1983a). Notes on new and noteworthy plants CLXIV. Phytologia 52: 414 – 415.

—— (1983b). Notes on new and noteworthy plants CLXXI. Phytologia 54: 399 – 400.

—— (1983c). Notes on new and noteworthy plants CLXIX. Phytologia 54: 66 – 68.

—— (1984a). Notes on new and noteworthy plants CLXXIV. Phytologia 55: 232 – 234.

—— (1984b). Notes on new and noteworthy plants CLXXIX. Phytologia 56: 380.

Múlgura de Romero, M. E., Rotman, A. D. & Atkins, S. (2003). *Verbenaceae*. Flora Fanerogámica Argentina. Fas. 84, No. 253 pt. 1. Proflora, Argentina.

Munir, A. A. (1992). A taxonomic revision of the genus *Stachytarpheta* Vahl (*Verbenaceae*) in Australia. J. Adelaide Bot. Gard. 14 (2): 133 – 168.

Nees von Esenbeck, C. G. & Martius, C. (1823). In: Maximilian von Wied-Neuwied, Beitrag zur Flora Brasiliens. Nova Acta Phys.-Med. Acad. Caes. Leop.-Carol. Nat. Cur. 11: 1 – 88.

Oliveira, P. S. & Leitão-Filho, H. F. (1987). Extrafloral Nectaries: their taxonomic distribution and abundance in the woody flora of *Cerrado* vegetation in Southeast Brazil. Biotropica 19 (2): 140 – 148.

Pires-O'Brien, M. J. (1997). Transverse dry belt of Brazil. In: S. D. Davis, V. H. Heywood, O. Herrera-MacBryde, J. Villa-Lobos & A. C. Hamilton (eds.), Centres of Plant Diversity. A Guide and Strategy for their Conservation. Vol. III The Americas: 397 – 404. IUCN.

Pohl, J. B. E. (1828). Plantarum Brasiliae icones et descriptiones 1. Vienna.

Raj, B. (1983). A contribution to the pollen morphology of *Verbenaceae*. Rev. Palaeobot. Palynol. 39: 343 – 422.

Ratter, J. A. & Dargie, T. C. D. (1992). An analysis of the floristic composition of 26 *cerrado* areas in Brazil. Edinburgh J. Bot. 49 (2): 235 – 250.

Richard, A. (1792). Catalogus plantarum, ad societatem, ineunte anno 1792, e Cayenna missarum a Domino le Blond. Actes Soc. Hist. Nat. Paris 1: 105 – 114.

Ridley, M. (1996). Evolution. Blackwell Scientific Publications, Cambridge.

Sales, M. F. *et al.* (1999). Composição Floristica e diversidade dos "brejos" de Pernambuco. In: F. D. Araujo, H. D. V. Prendergast & S. J. Mayo (eds.), Plantas do Nordeste. Anais do I Workshop Geral: 42 – 52. Royal Botanic Gardens, Kew.

Sanders, R. W. (2001). The genera of *Verbenaceae* in the south-eastern United States. Harvard Pap. Bot. 5 (2): 303 – 358.

Schauer, J. C. (1847). *Verbenaceae*. In: A. P. de Candolle, Prodromus 11: 522 – 570. Masson, Paris.

—— (1851). *Verbenaceae*. In: C. F. P. Martius (ed.), Flora Brasiliensis IX.

Schrader (1821). Gött. Gel. Anz. 1: 709.

Schultes, J. A. (1822). In: J. J. Roemer & J. A. Schultes, Mantissa 1. J. C. Cottae, Stuttgart.

Stannard, B. L. (ed.) (1995). Flora of the Pico das Almas. Royal Botanic Gardens, Kew.

Swofford, D. L. (2002)1. PAUP*: phylogenetic analysis using parsimony (*and other methods), vers. 4.0. Sinauer Associates Inc., Sunderland, Massachusetts, U.S.A.

Taylor, N. P. & Zappi, D. C. (2004). Cacti of eastern Brazil. Royal Botanic Gardens, Kew.

Turczaninow, P. K. N. S. (1863). *Verbenaceae* et *Myoporaceae* nonnullae hucusque indescriptae. Bull. Soc. Imp. Naturalistes Moscou 36 (2): 193 – 227.

Vahl, M. (1804). Enumeratio Plantarum 1. N. Mölleri et Filii, Copenhagen.

Verdcourt, B. (1992). *Verbenaceae*. In: R. M. Polhill (ed.), Flora of Tropical East Africa. A.A. Balkema, Rotterdam.

Walpers, W. G. (1845). Repertorium botanices systematicae 4 (1). Friderici Hofmeister, Leipzig.

Wurdack, J. J. (1970). Erroneous data in Glaziou collections of *Melastomataceae*. Taxon 19: 911 – 913.

Zappi, D. C., Lucas, E., Stannard, B. L., Nic Lughadha, E., Pirani, J. P., de Queiroz, L. P., Atkins, S., Hind, D. J. N., Giulietti, A. M., Harley, R. M. & de Carvalho, A. M. (2003). Lista das plantas vasculares de Catolés, Chapada Diamantina, Bahia, Brasil. Bol. Bot. Univ. São Paulo 21 (2): 345 – 398.

Appendix A

Characters used in the cladistic analysis:

Rootstock — present (0)/absent (1)
This is the presence or absence of an enlarged, very hard, woody rootstock at the base of the plant, just below ground level, and from which the roots descend. Coded as a simple presence/absence character.

Habit — Stems woody (0)/Stems herbaceous (1)
It is often difficult to know from herbarium specimens whether or not a plant is annual or perennial, therefore I have used this simple discrete character state.

Leaves — hairy (0)/glabrous (1)

Leaves — not coriaceous (0)/coriaceous (1)

Leaves — not rugose (0)/rugose (1)

Hairs — uniseriate (0)/simple (1)

Extrafloral nectaries — occurring on stem, leaves or calyx — present (0)/absent (1)

Inflorescence length — Up to 4 flowering nodes (0)/6 – 12 flowering nodes (1)/20 – 26 flowering nodes (2)/more than 35 flowering nodes (3)

Inflorescence width — Up to 5 mm (0)/8 – 15 mm (1)/18 – 30 mm (2)

Calyx position — Free (0)/Embedded (1)
The calyx is sometimes completely embedded in the rachis, and sometimes, only the very base is embedded. Where the whole of the calyx is embedded, the calyx is held in an erect position, parallel to the rachis. Where the calyx is not embedded, the calyx can be parallel with the rachis, or at an acute angle to it.

Calyx orientation — Erect — held parallel to the inflorescence axis (0)/Acute — forming an acute angle with the inflorescence axis (1)

Bracts — Woody (0)/Herbaceous (1)

Bracts — Persistent (0)/Caducous (1)

Calyx I — Posterior sinus present (0)/Posterior sinus absent (1)

Calyx II — Anterior sinus present (0)/Anterior sinus absent (1)

Calyx III — Posterior teeth expressed (0)/Posterior teeth not expressed (1)

Calyx IV — Anterior tooth expressed (0)/Anterior tooth not expressed (1)

Fruiting calyx — Not closed (0)/ Closed (1)

Corolla shape — Hypocrateriform (0)/Infundibular (1)

Corolla position — Erect (0)/ Acute (1)

Glandular hairs on outside of tube — Hairs present (0)/Hairs absent (1)

Corolla tube — Long (21 – 30 mm) (0)/Short (up to 20 mm) (1)

Corolla tube — With kink (0)/without kink (1)

Interior of tube — With hairs at throat (0)/without hairs at throat (1)

Interior of tube — With glands at throat (0)/without glands at throat (1)

Flower colour — White (0)/ White-pink (1)/ Blue (2)/ Red (3)/ Purple-black (4)/ Mauve (5)/ Yellow (6)

Number of fertile stamens — 4 (0)/2 (1)

Stachytarpheta has only 2 fertile stamens, but *Chascanum* and *Bouchea* have 4 fertile stamens.

Position of anthers — Top (0)/ Middle (1)

Anther thecae orientation — Divergent (0)/ Parallel (1)
Stachytapheta has divergent thecae, but *Chascanum* and *Bouchea* have parallel thecae.

Ovary — Not sitting on disk (0)/sitting on disk (1)

Stigma — Hooked (0)/Capitate (1). *Stachytarpheta* has a capitate stigma, *Chascanum* and *Bouchea* have a hooked stigma.

Surface of mericarp — Reticulate (0)/smooth (1), glandular (2)

Mericarp — With beak (0), without beak (1)

Apex of mericarp — Abscission point level with top of fruit (0)/ abscission point above level of fruit leaving short stylopodium (1)/abscission point below top of fruit, leaving v-shaped gap (2)

Appendix B

Taxa included in the morphological analysis

Section (Schauer)	Abena	
Subsection	Lepturae	S. polyura
		S. lactea
		S. schottiana
Subsection	Pachyurae	S. angustifolia
		S. lythrophylla
		S. jamaicensis
		S. microphylla

Section	Tarphostachys	
Subsection	Longispicatae	S. reticulata
		S. longispicata
		S. gesnerioides
		S. coccinea
		S. scaberrima
		S. quadrangula
Subsection	Brevispicatae	S. speciosa
		S. glabra
		S. trispicata
		S. martiana
		S. pachystachya
		S. hispida
Subsection	Subspicatae	S. commutata
		S. viscidula
		S. ajugifolia
		S. lychnitis
Subsection	Capitatae	S. lacunosa
		S. sellowiana
		S. discolor

Other taxa included were *S. almasensis*, *S. ganevii*, *S. arenaria*, *S. froesii*, *S. radlkoferiana*, *S. bromleyana*, *S. bicolor*, *S. caatingensis*, *S. amplexicaulis*, *S. cassiae*, *S. monachinoi*, *S. linearis*, *S. mexiae*, *S. procumbens*, *S. candida*, *S. dawsonii*, *S. sericea*, *S. atriflora*, *S. integrifolia*, *S. macedoi*, *S. crassifolia*, *S. tuberculata*, *S. stannardii*, *S. guedesii*, *S. cearensis* and *S. confertifolia*. Two taxa were included from Central America: *S. frantzii* and *S. mineacea*.

Appendix C

List of exsiccatae. Number in brackets following collector's number refers to taxon number in revision. Does not include specimens already cited in main text.

Agra & Gois 1224 (9). *Alves et al.* 4005 (61); 4025 (66); 4202 (19a); 4215 (19a); 4227 (19a); 4238 (17). *Amaral et al.* 675 (12a). *Amorim et al.* 1026 (19c). *Anderson et al.* 6460 (45a); 6661 (52); 6808 (77); 7109 (15); 7230 (1); 7410 (43); 7555 (77); 7672 (48); 7893 (45a); 8104 (45a); 8343 (68); 8515 (16a); 9672 (20a); 10021 (45a); 35436 (72); 35918 (67); 36961 (32). *Andrade et al.* 210 (12a); 238 (64). *Aparecida da Silva* 3093 (12); 3792 (50). *Arauja* 4415 (11); 4470 (11). *Arbo et al.* 3649 (50); 3704 (40); 4026 (68); 4088 (68); 4313 (68); 5036 (68); 5438 (19c); 5697 (36); 5770 (19a); 7182 (9); 7388 (26); 7548 (28); 7665 (75). *Atkins et al.* CFCR 13667 (66); CFCR 13680 (61); CFCR 13800 (68); CFCR 13932 (25); CFCR 14055 (19a); CFCR 14160 (19a); CFCR 14234 (60); CFCR 14258 (35); CFCR 14273 (19a); CFCR 14464 (60); CFCR 14586 (35); CFCR 14589 (3); CFCR 14686 (35); CFCR 14687 (62); CFCR 14790 (60); CFCR 14794 (60). *Avila* 15 (34).

Bamps 5501 (48). *Barros et al.* 2220 (45b). *Bautista et al.* PCD 4358 (60). *Belem* 1757 (9). *Belem & Mendes* 198 (30). *Bernacci* 974 (1); 1426 (?); 1801 (1). *Blanchet* 2820 (28); 3120 (26); 3126 (74); 3139a (?1); 3647 (19a); 3885 (32). *Brooks et al.* 157 (45a). *Burchell* 6205 (45a); 6340 (2); 7990 (77); 8029 (43); 9071 (12a)

Carneiro & Oliveira 91 (19a). *de Carvalho* 277 (1); 539 (29); 2285 (43); 2698 (75); 2954 (19a); 3035 (28); 3246 (32); 3696 (60); 3778 (32); PCD 983 (37); PCD 1006 (19a). *Castellani et al.* 175 (1). *Chukr* 8 (1). *Claussen* 131 (58); 205 (66); 386 (21). *Clayton* 4922 (45b). *Cobra & Sucre* 401 (49). *Collares & Fernandez* 134 (47). *Collares et al.* 138 (27). *Collenette* 187 (20a). *Coradin et al.* 6211 (74); 6390 (36); 6499 (19a). *Cordeiro et al.* SPF 22864/CFCR 937 (16b); SPF 22885/CFCR 958 (73). *Cordeiro & Kummrow* 202 (1). *Costa* 115 (1). *Costa & Assunção* 763 (1). *Culau* 21 (6). *Custodio filho* 470 (1). *Czermak* 70 (7)

Damasceno Junior 1114 (6); 2041 (6); 2956 (12). *Dick* 43 (1). *Dombrowski* 12329 (1). *Dorrien Smith* 239 (20a). *Drouet* 2218 (12a). *Dubs* 1415 (5); 2073 (5)

Eiten 7001 (68). *Eiten & Eiten* 3776 (27); 10131 (1). *Emygdio et al.* 3275 (59). *Esteves & Barros* 2581 (64)

Farias 429 (9). *Ferreira, L.V.* 29 (45b). *Ferreira, M.P.* 4 (45b); 644 (70). *Figueiredo et al.* 151 (64). *Folli* 1193 (9); 2679 (1). *da Fonseca* 410 (31). *Forero et al.* 8410 (1). *Forzza et al.* 375 (45b). *Fothergill* 65 (60); 67 (39); 131 (19b). *França & Melo* 1058 (29); 1124 (26); 1184 (12a). *França et al.* 995 (19b); 1273 (19b); 4188 (75); 4342 (30); 4344 (30); 4347 (16b); 4349 (16b); 4370 (73); 4391 (1); 4429 (64); 4461 (57); 4464 (57); 4472 (68); 4477 (68); 4505 (67); 4539 (14); 4543 (69); 4549 (16a); 4557 (63); 4565 (63); 4571 (68); 4599 (25); 4600 (45b); 4618 (43); 4622 (43); 4654 (77); 4658 (43). *Franco* 1313 (1). *Freire-Fierro et*

al. 1655 (19a); 2064 (19a). *Freitas 2* (19a). *Froes* 20140 (37). *Furlan et al.* SPF 18013/CFCR 418 (19a); SPF 18842/CFCR 2047 (36); SPF 18966/CFCR 1589 (19a); SPF 20295/CFCR 7488 (53); SPF 22685/CFCR 751 (16b); SPF 23281/CFCR 3074 (63); SPF 37190/CFCR 7390 (60)

Ganev 157 (35); 209 (28); 333 (40); 500 (62); 668 (40); 748 (28); 1576 (28); 1804 (35); 1860 (60); 1887 (62); 2066 (19b); 2249 (19b); 2566 (3); 2737 (19b); 2869 (28); 2924 (19b); 2935 (19b); 2973 (3); 3083 (35); 3171 (19b); 3336 (19b); 3383 (19a); 3421 (35). *Gardner* 831 (32); 1105 (4); 1106 (12a); 1383 (10); 1815 (27); 1816 (1); 1838 (12a); 2285 (13); 2286 (26); 2434 (32); 3015 (19b); 3410 (77); 3934 (12a); 3935 (77); 3936 (43); 4337 (73); 4338 (?); 4339 (47); 5109 (25); 5110 (68); 5111 (21); 5112 (21); 5113 (59). *Gasson et al.* PCD 6179 (32). *Gates & Estabrook* 18 (48); 176 (45a). *Giulietti et al.* 1583 (28); 1701 (75); PCD 799 (60); PCD 3248 (19); SPF 18103/ CFCR 1240 (36); SPF 18463/CFCR 1551 (35b); SPF 21616/CFCR 1835 (69); SPF 36373/CFCR 6773 (3). *Glaziou* 11326 (27); 11327 (1); 13054 (10); 13061 (76); 13062 (12a); 13063 (22); 14161 (76); 15328 (68); 15329 (57); 17714a (25); 18392a (20a); 19718 (77); 19719 (21); 19720 (64); 19721 (64); 21893 (54); 21894 (20a); 21903 (48); 21904 (48); 21905 (46); 21906 (43); 21907 (46); 21908 (49); 21909 (45b). *Godoi & Neto* 182 (1). *Gonçalves* 127 (74). *Gottsberger* 11-30168 (1); 12-15186 (12a); 12-23768 (8); 17-251085 (21). *Gouveia* 43/83 (31). *Guedes et al.* PCD 482 (60); PCD 699 (19a); 932 (4); PCD 5524 (60); PCD 5666 (60)

Harley et al. PCD 3277 (74); PCD 4476 (60); PCD 4996 (60); SPF 35942/CFCR 6236 (68); SPF 36078/CFCR 6478 (69); 11362 (51); 11470 (43); 15136 (60); 15533 (36); 15853a (36); 15896 (19a); 15938 (35); 16840 (74); 16849 (60); 17031 (1); 17092 (9); 17555 (1); 17923 (10); 18295 (9); 18597 (28); 18668 (19a); 19196 (19c); 19297 (19c); 19454 (29); 19655 (60); 19683 (19b); 19940 (60); 20035 (3); 20183 (3); 20526 (29); 20558 (28); 20658 (19a); 20702 (36); 20797 (19a); 20852 (3); 20861 (38); 20936 (3); 20964 (28); 20981 (35); 21124 (60); 21264 (60); 21477a (15); 21579 (31); 21692 (31); 21868 (1); 22456 (19a); 22489 (37); 22667 (19a); 22831 (19c); 22942 (60); 24173 (35); 24264 (60); 24488 (19b); 24530 (60); 24922 (21); 25009 (21); 25058 (73); 25060 (16b); 25235 (19d); 25421 (59); 25730 (60); 26012 (60); 26114 (62); 26430 (19b); 26467 (36); 26505 (60); 26559 (39); 27335 (3); H50246 (19b); H50525 (19b); H50649 (35); H50650 (18); H50681 (60); H50846 (35); H51291 (35); H51996 (3); H52088 (60). *Hatschbach et al.* 29098 (14); 29523 (6); 29961 (53); 30115 (67); 30152 (21); 31674 (21); 31700 (14); 36361 (55); 37248 (54); 42090 (13); 42347 (74); 44170 (26); 44687 (63); 46311 (22); 46323 (23); 46386 (75); 46523 (62); 47023 (9); 47464 (19c); 47946 (35); 48359 (36); 50074 (28); 50080 (75); 50954 (67); 52035 (1); 53839 (43); 54662 (50); 55932 (45a); 56284 (50); 56659 (12a); 56855 (19a); 56858 (35); 58339 (43); 59296 (45a); 59332 (48); 60021 (45c); 60787 (6); 61814 (21); 61952 (28); 62024 (60); 63050 (9); 63053 (9); 63127 (12a); 64234 (21); 64318 (69); 64676 (21); 64789 (45d); 65321 (1); 66023 (28); 66191

(21); 66320 (69); 66420 (21); 66436 (68); 66476 (16a); 66884 (20a); 67231 (53); 67594 (60); 67727 (74); 67899 (60); 67985 (16b); 68489 (9); 68499 (9); 69003 (21); 69210 (8); 69535 (53); 69712 (63); 69752 (68); 70051 (45b); 70422 (43); 70643 (48); 70677 (45a); 70751 (43); 71057 (73); 71069 (77); 71128 (12); 71139 (45); 71540 (8); 71991 (21); 72033 (45d); 72073 (69); 72256 (21); 72942 (1); 73198 (1); 74113 (1); 74577 (12); 74848 (12); 74990 (1); 75273 (9); 75509 (26); 75828 (29); 75834 (12). *Henrique & Kawasaki* SPF 20430/CFCR 7640 (53). *Hensold et al.* SPF 21916/CFCR 2667 (68); SPF 22305/CFCR 2825 (68); SPF 23347/CFCR 3139 (68); SPF 30083/CFCR 3504 (73). *Heringer* 8179/373 (43). *Heringer et al.* 1053 (20a); 2915 (43); 6433 (43). *Hill, S.* 13083 (1). *Hind et al.* 050 (9); PCD 3170 (19c); PCD 3518b (17); PCD 3555 (19a); PCD 4259 (3); PCD 4566 (19a); PCD 4567 (28); H51405 (32). *Hoehne* 5648 (7). *Hora & Campelo* 72 (1). *Hunt et al.* 5539 (20a)

Irwin et al. 7801 (43); 7822 (45b); 8326 (2); 8706 (43); 9583 (43); 10721 (45b); 11292 (1); 12393 (54); 12421 (54); 12682 (3); 13232 (48); 13774 (45a); 14170 (43); 14455 (47); 15104 (12b); 19611 (68); 20224 (59); 22157 (21); 22625 (68); 23029 (21); 23425 (73); 23871 (21); 24136 (43); 24277 (52); 24675 (43); 24904 (48); 24941 (52); 25165 (2); 25457 (45c); 26462 (43); 26918 (20a); 25016 (20a); 27051 (21); 27702 (68); 28208 (21); 28989 (68); 29635 (68); 30692 (19c); 32004 (48); 32857 (55); 33116 (52); 33117 (45a); 34652 (20a); 48795 (1)

Joly et al. 340 (1); 1913 (69)

Kawasaki et al. SPF 36194; CFCR 6594 (57). *Kirkbride* 1145 (1); 4852 (43). *Krapovickas & Cristobal* 33180 (45b); 33188 (43); 33418 (25). *Kuhlmann* 1834 (1)

Læssøe & Sano H52335 (35). *Landrum* 4170 (11). *Lasseigne* P22596 (1). *Leitão Filho et al.* 12300 (20a); 34499 (1); 34695 (1). *Leite et al.* 153 (26). *Lewis* SPF 36477/CFCR 7477 (12a). *Lima* 53-1244 (1); 53-1472 (12a); 58-2958 (45b); 58-3008 (49); 58-3098 (20a). *Lima, L.* 211 (74). *Lindman* A607 (7). *Löfgren* 160 (27); 692 (32). *Longhi-Wagner et al.* SPF 34987/CFCR 5927 (25). *Lowe* 4172 (1). *Luetzelburg* 214 (35); 253 (39); 26208 (64)

Macedo 1647 (15). *Magalhaes* 2629 (53); 2736 (59); 6024 (56). *Makino et al.* 25 (1). *Mamede et al.* SPF 23593/CFCR 3388 (16b). *Mantovani* 292 (1); 426 (1); 1291 (1); 1528 (20a). *Martinelli & Hatton* 10083 (1). *Martinelli et al.* 5362 (19a). *Martins* 1027 (68). *Martius* 1044 (58). *Matiko Sano & Filgueiras* 47 (45a). *Mattos, J.* 8643 (1); 13031 (1); 15673 (1). *Mello-Silva et al.* 436a (19d); 492 (68); 558 (48); 694 (16b); 801 (28). *Melo et al.* 937 (9); PCD 1177 (37); 1196 (17); PCD 1248 (19a). *Melo & Chiea* 201 (1). *Mendes* 3421 (1). *Mendonça et al.* 2428 (45a). *Menezes et al.* SPF 18355/CFCR 1443 (35); SPF 20124/CFCR 7299 (53); SPF 21445/CFCR 545 (68); SPF 21658/CFCR 142 (21); SPF 22548/CFCR 613 (45a). *Mexia* 5753 (68); 5824 (59). *Moldenke & Moldenke* 19674 (7); 19996 (11); 19999 (11). *Mori & Boom* 14157 (26); 14371 (37); 14389 (19a).

Mori et al. 9992 (32); 10892 (9); 11102 (32); 11743 (1). *de Moura* 639 (12a)

Nee 42373 (1); 42951 (1). *Neto et al.* 1398 (1). *Nic Lughadha* PCD 6109 (32); H52011 (40); H52396 (40). *Nonato et al.* 908 (19a). *Nunes et al.* 105 (19b)

de Oliveira 27 (1); 413 (29). *Oliveira A. et al.* 97 (19a). *Onishi et al.* 877 (20a). *Orlandi* 392 (32)

Paciornik 161 (5). *Passos et al.* PCD 5110 (26); PCD 5399 (74); PCD 5775 (61). *Peateado* 1 (34). *Pereira, E.* 8975 (48). *Pereira & Alvarenga* 2992 (73). *Pereira & Pabst* 8643 (31); 9450 (11); 9700 (32); 9768 (75). *Pereira et al.* PCD 264 (19a). *Philcox* 3980 (1). *Philcox & Fereira* 3788 (1); 3872 (20a); 4019 (1); 4479 (1). *Philcox & Onishi* 4309 (45b); 4336 (1); 4767 (45b); 4914 (43). *Pirani et al.* 1523 (45a); 1613 (45a); 1650 (45b); 1818 (52); 1859 (54); 1901 (54); 1922 (12a); H51385 (28). *Pires & Cavalcanti* 52359 (12a). *Plowman et al.* 12069 (12a). *Prance et al.* 12528 (1)

de Queiroz 296 (75); 594 (19a); 633 (37); 2792 (1); 5031 (19b); 5529 (17); 7250 (26); 7294 (75); 7605 (69); 7803 (29); 7894 (74).

Rabelo 3308 (1). *Ratter et al.* 910 (20a); 1372 (20a); 2775 (45b); 2942 (20a); 3657 (1); 7267 (51); 7271 (45a); 7278 (55); 7291 (52). *Ribas & Hirai* 2008 (1). *Ribas & Pereira* 1717 (2); 1838 (1); 2579 (2). *Rodal* 262 (64). *Roque et al.* PCD 2331 (19c); CFCR 15004 (60); CFCR 15014 (78); CFCR 15179 (21); CFCR 15193 (20a); CFCR 15302 (53); CFCR 15308 (21). *Rosado* 18 (45b). *Rossi et al.* SPF 19995/CFCR 7272 (70); SPF 22968/CFCR 1042 (21); SPF 23535/CFCR 3328 (69)

Saar et al. PCD 5029 (28). *Saint-Hilaire* 2941 (66). *Sales* 429 (64). *Sano, P.T.* CFCR 14857 (60). *Sano & Læssøe* H50861 (18); H50993 (19b). *de Sant'Ana et al.* 288 (1). *dos Santos, E.B.* 12 (12a). *dos Santos T.S.* 3263 (9). *Santos J.U.* 503 (20b). *Santos T.R. et al.* 5852 (35). *Schwacke* 9406 (66); 11530 (57); 10104 (65); 10105 (65); 12346 (66); 13465 (66). *Segadas-Vianna et al.* 975 (8). *Sellow* 162 (30); 223 (11); 552 (19a); 950 (65); 1456 (70); 5468 (20a). *Silva, S.M.* 1157 (20). *Silva-Castro et al.* 657 (32). *Smith, L.B.* 1420 (1). *Sobral & Mattos Silva* 5897 (31). *Souza, J.P. et al.* 10 (1). *Souza, V.C. et al.* 5305 (28); 5443 (31); 11000 (1). *Stannard et al.* 1001 (8); PCD 2842 (19a); PCD 5237 (32); SPF 35772/CFCR 6122 (14); SPF 35778/CFCR 6178 (16a); H50835 (18); H51062 (56); H51065 (18); H51242 (18); H51901 (59); H51983 (19b); H51996 (3)

Taylor et al. 1465 (32). *Teixeira, I.M.* 5 (45b). *Thomas W.W. et al.* 5915 (20b); 5990 (30); 6174 (11). *Tozzi & Teixeira* 94-253 (1); 94-252 (1)

Ule 5411 (12a); 7414 (26); 7866 (12a); 7975 (24); 8295 (24). *Ussui et al.* 15 (1); 33 (34)

Villarouco 4 (64). *da Vinha & Castellanos* 21 (1). *da Vinha & Santos* 147 (9)

Webster et al. 25752 (74). *Weddell* 1534 (25). *Williams & Assis* 6639 (55); 7920 (59)

Yale Dawson 14209 (77); 14288 (48); 14534 (1); 14618 (45a); 14659 (43); 14722 (47); 14758 (1); 14981 (20); 15053 (50); 15058 (20)

Index

(accepted names in **bold**, synonyms in *italics*)

S. microphylla *Walp.*	26	S. scaberrima *Cham.*	30
S. *mollis* Moldenke	= 43	var. *pilosa* Moldenke	= 9
S. monachinoi *Moldenke*	56	S. *schaueri* Moldenke	= 43
S. *obovata* Hayek	= 73	S. schottiana *Schauer*	11
S. pachystachya *Mart. ex Schauer*	77	var. *angustifolia* Moldenke	= 11
S. paraguariensis *Moldenke*	5	S. sellowiana *Schauer*	65
S. piranii *S. Atkins*	42	S. sericea *S. Atkins*	48
S. pohliana *Cham.*	70	S. sessilis *Moldenke*	27
S. polyura *Schauer*	2	S. *simplex* Hayek	see notes at end
f. *albiflora* Moldenke	= 3	S. spathulata *Moldenke*	16
S. procumbens *Moldenke*	53	subsp. **mogolensis** *S. Atkins*	16b
S. *prostrata* Loes. ex Glaz.	= 54	S. speciosa *Pohl ex Schauer*	34
S. puberula (*Moldenke*) *Atkins*	44	S. sprucei *Moldenke*	24
S. quadrangula *Nees & Mart.*	28	S. stannardii *S. Atkins*	18
S. radlkoferiana *Mansf.*	35	S. *subulata* Moldenke	see notes at end
var. **lanata** *S. Atkins*	35b	S. *triphylla* (Pohl) *Walp.*	= 49
S. restingensis *Moldenke*	8	S. trispicata *Nees & Mart.*	31
var. *hispidula* Moldenke	= 11	var. *ovatifolia* Moldenke	= 31
S. reticulata *Mart. ex Schauer*	21	S. tuberculata *S. Atkins*	17
var. *bahiensis* Moldenke	= 62	S. villosa (*Pohl*) *Chamisso*	43
S. rhomboidalis (*Pohl*) *Walpers*	49	S. *villosa* Chamisso	= 43
var. *puberula* Moldenke	= 44	var. *bahiensis* Moldenke	= 75
S. rotundifolia *Link*	see notes at end	S. viscidula *Schauer*	58
S. rupestris *S. Atkins*	25	var. *brevipilosa* Moldenke	= 67
S. *sanguinea* Schauer & Mart.	= 26		
var. *grisea* Moldenke	= 32		
var. *hatschbachii* Moldenke	= 26		